TOMORROW'S WAR

Original title: *La guerre probable – Penser autrement,* 2e édition, Economica, 2008

Translation: Joanna MacDaniel

The "*Stratégies & Doctrines*" Series

General Vincent DESPORTES

TOMORROW'S WAR

Thinking otherwise

Preface by General (U.S.) **William S. WALLACE**

ЄӘ **ECONOMICA**
49, rue Héricart, 75015 Paris

Also by Vincent Desportes

Cavalerie de décision – Pérennités et perspectives, ADDIM, 1998

Comprendre la guerre, Economica, 2000
(Prix de l'Académie des Sciences Morales et Politiques, Prix Vauban de l'IHEDN)

L'Amérique en armes – Anatomie d'une puissance militaire, Economica, 2002

Décider dans l'incertitude, Economica, 2004

Introduction à la stratégie, Economica, 2007
(with Jean-François Phélizon)

Deciding in the Dark, Economica, 2008

La guerre probable – Penser autrement, 2ᵉ éd., Economica, 2008

This book is dedicated to those who, in these uncertain times, will bear the heavy responsibility of deciding on our force models for the future: our politicians.

ACKNOWLEDGEMENTS

I would like to thank those authors whose innovative thinking has ena-ble me to grasp the dramatic changes currently taking place with regard to conflict and the conditions under which our forces are engaged, in partic-ular, Bertrand Badie, Jacques Baud, Colin S. Gray, Frederick W. Kagan, Herfried Münkler and Sir Rupert Smith.

I would also like to thank Caroline Galactéros, Philippe Prevost and Luc de Revel, who kindly agreed to read my manuscript and who offered such helpful advice.

The first, the supreme, the most decisive act of judgment that the statesman and commander have to make is to establish the kind of war on which they are embarking; neither mistaking it for, nor trying to turn it into something that is alien to its nature.

Clausewitz

PREFACE

Globally, we are faced with an era of "persistent conflict" in which our Nations will be required to conduct full spectrum operations against both regular and irregular threats. The aging debate of how we must adapt to win today's fight while preparing for future conflict has never been more pronounced while the need to balance future capabilities with those needed for today's conflicts never more urgent.

I have been intimately involved in this debate since my time as the Commander, U.S. Army V Corps in Iraq. I have had the unique opportunity to fight in the environment, as well as, study, discuss, and at times been charged to lead the U.S. Army's thoughts and actions in this debate. It was during this time, that I was fortunate enough to candidly discuss ideas with MG Vincent Desportes. We share many of the same beliefs. Chief among these beliefs is an understanding that finding the final answers, adapting the military institution and changing political thought requires careful thought, negotiation and understanding. In Desportes'book, *Tomorrow's War – Thinking Otherwise,* the reader gets just that from a veteran insider to the debate. MG Desportes combines a senior leader's depth of thought and understanding with an uncanny ability to convey his ideas in a way both Soldiers and Statesmen can understand and readily apply to the future.

Readers will be intrigued as MG Desportes examines the global trends and changing operational environment leading to *why* collectively our nations and armies must adapt. But perhaps more important than *why*, the reader will be challenged to question his/her own thinking as MG Deportes provides an expert's viewpoint of *how* to adapt our very way of thinking and operating – politically and militarily, while building and employing useful military power in this new environment.

His ideas provide a blueprint for *how* to adapt our thinking and operational concept. He emphasizes a future where conflict resolution will require a protracted commitment of all elements of national power and battlefield success by itself will no longer be enough.

Additionally, MG Desportes devotes considerable thought in **how** to build useful military power in this new environment. His critical analysis of the U.S. Army's transformation process provides a candid look at our tendencies to focus on technology and conventional power while mortgaging the ability to win today's war and succeed in the most likely future conflict scenarios. He astutely sums it up by providing his provocative ideas on applying these lessons learned to the future force.

Not all will agree with all of his ideas. However, all leaders, military and civilian alike, will benefit from this deep look into **why** and **how** our coalition of nations and armies must adapt for success in the future operational environment. One needs to look no further than the U.S. Army's recent publication of FM 3-0, *Operations*, FM 3-07, *Stability Operations*, and FM 7-0, *Training for Full Spectrum Operations* to see evidence of our many shared beliefs. In the end, our enduring success will only be achieved through our commonly agreed upon resolution of this debate, shared responsibility – politically and militarily, and a persistent commitment by all elements of national power.

William S. Wallace

General, U.S. Army Retired

GENERAL INTRODUCTION

As idealists and the children of enlightenment and civilisation, we like to think that war is dead.

Alas, this hope is just as much a part of mankind as war itself. Since the dawn of humanity, man and war have been inseparable. Man consists of desires –a desire to live and a desire to dominate– and confrontation is an inextricable aspect of the meeting of these desires.

The last war, no matter how bloody, is simply another milestone in the history of humanity, which is, at the same time, the history of war. Neither the League of Nations, that still-born child, nor the United Nations, the child of renewed hope, were able to end war, for while war destroys, it does not die. War did not die in Versailles in 1919, any more than in San Francisco in 1946, or as a result of the hope born of the nuclear deterrent, or indeed with the fall of the Berlin Wall.

On the contrary, having been contained for some time, it is now spreading and gaining strength. Old slumbering quarrels, new desires for power and the basic need to survive are combining with fundamentalism and the multiplier effects of globalisation to give conflicts and violence a new force, which is further intensified by the new transparency of the modern world.

War is returning in force, accompanied by re-arming. The planet is in a constant state of re-armament, with current world military spending back at the levels of the Cold War. It is back in force, but while its nature may not have changed, its face most certainly has.

War has a magnificent past; it also has a glorious future.

WE MUST PREPARE OURSELVES FOR WAR

War is our past, our present and also our future. We have to be prepared. Of course, the armed confrontations in the outside world do not necessarily involve us, but to doze off in our artificially secure bubble would be to leave ourselves open to a rude awakening when tomorrow, or the day after, in some almost unpredictable way, war returns to our doorstep. Let there be no doubt, if you take no interest in what seems a distant crisis, the crisis will be very likely to be interested in you. If we simply sit back on our balconies and observe war uncritically, violence, currently contained outside our door, will soon cross the threshold. In the past, we have mastered the uncertain threat, but if we do not treat it with respect, the threatening uncertainty of today will become tomorrow's reality; it will make its presence felt in our land and on our streets, right there amongst us. So to avoid being consumed by the inferno, the only solution is to tackle it head on, working first to contain, then to reduce and finally to eradicate the root of the violence. In all cases shutting our eyes to what is happening will lead to failure, all firm defences will collapse eventually and there is no Maginot Line that cannot one day be bypassed.

If the protection of territory, both national and overseas, our fellow citizens and our essential values is considered important –and, in principle, nothing should be more important to a state– our civil and military leaders should concern themselves with conflict in its new forms. They should endeavour to understand it and to grasp its development, in order to enable France to act anywhere in the world to provide this protection. However, in order for any plans for political intervention to be suited to the new contexts, and for the armed forces to be able to execute their missions to achieve these political aims, it is important to understand the evolution that war has undergone and to know what form war will be most likely take. There is a clear relationship between the way a country is listened to and its useful military strength. In order to help the voice of France to be heard across the world, it is important to examine the models of available force and of military effectiveness in tomorrow's warfare.

The response to such examination and questioning is, however, not an easy one. Current state structures, analysis matrices, equipment production processes, armaments industries and military apparatus are all largely the legacy of the paradigms of the Cold War. In the absence of concrete proof of their obsolescence, the concepts on which they are based seem to retain a reassuring virtue, whereas the strength of what we possess –the arsenals put together at great cost– considerably restricts how far we are prepared to alter our views and our military systems.

Thus we cannot allow ourselves the economies that would be achieved by a vigorous examination of the new conditions governing the use of force or of changes in military effectiveness and its associated developments. Do the paths which we chose to follow then, and which we are essentially still following, still meet our requirements? The answer is probably no, which is by no means illogical. In the past, our vision of the world and of war led to appropriate modes of action. At the same time, the rule of circumventing an obstacle, which is really the first rule of war, led the potential opponent to change, and to seek responses and new forms of resistance to the new weapons we had invented. There are very few opponents who would wish to become the consenting victim of the sort of warfighting we are familiar with, that we prefer to conduct and for which, in many instances, we are still preparing.

A GRADUAL RAISING OF AWARENESS

Although the strategic transformation of space and time resulting from the effects of the internationalist erosion of the Westphalian order and the intrusion of the rapid pace of electronics had been apparent for some time, 11 September 2001 marked a watershed. It has often been said, but it is still important to be aware that these dramatic events truly crystallised the changes in the use of force that had gradually been taking place.

The collapse of the Soviet empire initially gave rise to a feeling of omnipotent euphoria and to the certainty that the constituents of its success made it possible for the West to do anything it wanted. Thus, able to do everything, it had to do everything. The decade of the "right to interfere" arose from the combination of this moral duty and the certainty that the classical attributes of power –as they had been perceived for the previous fifty years– made it possible everywhere to impose the true order and the true moralality, namely ours.

Sadly, the horizon has been steadily darkening. The political effectiveness of our military strength has, step by step, been called into question. The situation in the Balkans and the inability of our forces to settle the disputes there easily should have been an indication that something had changed. Nevertheless, for a long time we believed that these conflicts were the exception, whereas they were in truth the rule of a new reality. We simply made semantic changes, referring to these crises as "operations other than war" and by distinguishing, wrongly, between operations aimed at coercion and those aimed at controlling violence.

It took Afghanistan, then Iraq, and then Lebanon for us to become fully aware that the reason our military strength was unable to achieve the desired political aim was because the model of war had changed, and that applying force designed for a political context which had disappeared, according to methods designed for that same context, would be of no use.

War had changed its appearance and the constant evolution of circumstances rendered adaptability an essential quality for armies facing opponents more skilled at innovation.

We have created an expeditionary-style force based, for reasons of logistic necessity, on density and speed, for brutal and quick campaigns, in the certainty that the speed of the decisional loop alone would be sufficient for victory. This took no account of Clausewitz's law of reciprocal actions and treated the opponent as if he were just like us, in other words, with no regard as to his adaptability! This displayed a lack of understanding of the fact that, when faced with an opponent who refuses to "play the game" and circumvents strength and its procedures, or an opponent who decides to avoid our "chronostrategies"[1] by voluntarily accepting the long-term approach, the tool we have devised is capable of acting increasingly rapidly. Increasingly often, this comes also at the wrong moment or has the wrong effect. It also took no account of the fact that the political impatience of our democracies makes it difficult for them to act in areas where time is measured in the "long term", and in pursuit of long-term social and political objectives. It further failed to understand that speed, in itself, can be counter-productive, as it makes it difficult to understand the situation and, even more so, to make the necessary adaptations. It is based on the false premise that a smaller army, reduced as a result of the peace dividend and the focus on the paradigm of a high degree of stand-off precision, would be suited to the 21st century and its constantly changing forms of conflict. It failed to grasp the idea that, paradoxically, the expeditionary nature of the forces made them unsuited to the new types of expeditions.

We have, however, gradually understood that achieving the desired political effect requires far more than a simple tactical and technical victory, more than a local effect. We have also noted the transition from the old, industrial form of war to wars fought among the population, where any operation, large or small, must first and foremost be seen as a communications operation. We have seen that it is societies and people which form the new environment of action. We have also seen, further down the line, the consequences of these developments in terms of the design and use of weapons, and coordination with non-military actors. We have rediscovered the urgent need to be "control the environment", in other words to be present in the theatre, in large numbers, on a long-term basis. We have understood that, contrary to a widespread misunderstanding, numbers are in themselves an important characteristic which, for numerous reasons, is crucial to both effectiveness and protection. We have seen that

---

1. Henrotin.

while firepower is indispensable, it cannot compensate for a shortage of manpower. We have also realised that it is much more important to "understand" than to "know". We have learned that it is during the stabilisation phase –the decisive phase of the operation– that the conditions needed for strategic success are created, and that the de-escalation of violence, help for the civilian population and re-establishing their "social contract" are essential. We have grasped the fact while the use of force alone can no longer be the foundation on which military effectiveness is based, the military must be in possession of the best possible tools. As the ultimate insurance in the event that we are unable to prevent the re-emergence of a major threat, in future wars these tools will be crucial to the success of the intervention phase. They will also play a major role in terms of deterrence and, if necessary, coercion.

THINKING OTHERWISE

History shows us that, if a state wishes to be able to influence the destiny of the nation and to play a significant role in the worldit must have not simply a military force but a useful military force. The feeble performance of our current tools obliges us to understand the evolution of forms of strength, of nuisance and of violence. This is a matter of survival, for our civilisation and for ourselves. Our world is a finite world, with no room to rebuild destroyed hopes or lost terrain. There are no new worlds and so there are no new solutions, other than to preserve and manage together the world in its present form, from which it is impossible to escape. The future lies within us and in our powers of understanding and adaptation, nowhere else. We, the political and military leaders, thus have to give force back its utility; this force which is as essential today as it was yesterday. Restoring utility to force means understanding wars we will probably be called upon to fight and the political factors involved. We need to reflect long and hard and to act on the conclusions.

We need to change our thinking.

PART ONE

RETURN TO HISTORY

"Rising up to see better, Correlating facts to understand better, Identifying situations to act better."

Joël de Rosnay

CHAPTER I

USEFUL MILITARY STRENGTH

Over the last fifteen years, the classic form of military engagements has declined to such an extent that the dimensions of strength and the usable form of the associated instruments have altered dramatically; this leads to a need to take a fresh look at history and the often neglected past. Inter-state violence has not disappeared but, having lost its predominance, it is now neither the most likely nor the most dangerous threat to Western countries, given the considerable imbalance in the weapons available.

In this new framework, which is gradually being confirmed and solidly established, conflict has lost its traditional identity. The notion of strength has been replaced by the notion of sense and the classic military tools, forged using the latest technology for use against states whose very legitimacy is now weakening, are become increasingly less relevant.

Military strength today is not what it was.

GETTING OUT OF A TIGHT SPOT

On leaving his post as Chief of Staff of the Israeli Defence Forces, General Dan Halutz observed, that the growing complexity of asymmetric threats made it necessary to rethink what one could expect from deploying armed forces. Given the considerable difficulty the Israelis had experienced in reaching their objective in the war against Hezbollah in July/August 2006, he also questioned the concepts of "decisive action" and "victory". In his view, the meaning of these two concepts has changed appreciably. He also expressed concern about the suitability of resources to achieve the desired aims. And why shouldn't he?

Confronted by Hezbollah's 10,000 men, the Israeli Army had deployed the equivalent of the French land and air forces in the immediate vicinity of its bases. It fired between 40,000 and 100,000 tank or artillery rounds, launched over 2,000 missiles and flew over 15,000 sorties (of which two thirds were attack missions), using some of the most advanced weapons in the world. Despite all this, Israel was unable to achieve its desired aims; at no time did the attacks, from land and from the air, reduce Hezbollah's ability to command, erode their forces' fighting capability, or stop rockets being fired at Israeli territory. Over 4,000 rockets[1] were fired during this war which was intended to put an end to these attacks. However, the cost of this non-victory is considerable. For the 600 Hezbollah combatants killed[2], and immediately replaced, Israel lost 120 men and spent 6 billion dollars, to which one should add the 7 billion dollars worth of damage caused in Lebanon. In addition to these astounding sums, one must not forget the costs relating to the one million displaced persons (amounting to a quarter of the Lebanese population) and the 1,200 civilians killed and 3,700 wounded.

The questions raised by General Halutz are fundamental. Their answers form the basis of the aims of military action and thus of the way they are implemented: doctrines, force models, instruction, training. What are the aims of the military action, what are the limits on the use of force, and how should force be adapted to ensure economic and strategic effectiveness despite the clear evolution of the context of engagement? Deep down, what General Halutz was questioning in the face of this Israeli powerlessness, was the very concept of power, or even that of effective military power.

Such questions, while worrying, are not surprising. The American engagement in Iraq should have been a time for reflection and change. Since 1 May 2003, and the somewhat hasty statement by President Bush on the aircraft carrier USS Abraham Lincoln, the observer has been drawn into the sad and convoluted history of this military engagement, which seems to get more and more entangled, lost in yet another tale of the meddler being caught in his own meddling. The growing list of attacks, the increasing numbers of coalition losses, the daily recitation of the numbers of bodies found each morning in the gutters of Baghdad, the inability to achieve the defined aims and the worldwide decline in the security situation radiating from the Iraqi carbuncle have ultimately created a feeling of perplexity and concern, perhaps even anguish, in the face of the collapse of models previously thought to be sound. The rapid victory achieved has been replaced by an infernal logic of actions and reactions, with the coalition troops gradually sinking into a horrendous mire and the Iraqi people once again torn apart in what has become a particularly painful episode of history.

---

1. The figures given here are principally those used by the CSIS, Washington, USA.
2. Israeli figure: Hezbollah gave a figure of 70, which would increase tenfold the relative costs calculated.

The uncertain future of this desperate situation presents a serious challenge to classic strength and its resources. Summing up this concern, Arnaud Blin writes: "After the interventions in Afghanistan and Iraq we see, with a considerable degree of incredulity, that the United States were incapable of imposing their will in two 'second-class' theatres. In particular, we note that the vast sums invested in considering future strategy did not prevent the government from completely misjudging the nature of the stakes and the adaptability of contemporary strategic choices. This makes the debate about the RMA (Revolution in Military Affairs) seem all the more sterile." Thus, the disruption of world order is accelerating in the laboratory of Iraqi ambushes, while the customary and reassuring array of instruments of power is being turned on its head. The worried observer notes that the build-up of technical capacity may, if care is not taken, turn out to be nothing more than a build-up of political impotence. In the bloodbath of the suicide attacks, war is moving further and further away from the simple model of a technical confrontation between two arsenals, while the correlation between strength and the ability to seize the advantage –to produce strategic effect– is becoming far less clear-cut. Each tragic day confirms that, with regard to the armed forces, their political substance far outweighs their technical capabilities!

Immediately after the brilliant campaign of March/April 2003, instead of once again epitomising Western military superiority, this war –rather than being an "operation other than war", as the Americans put it– became "an operation worse than war". This clearly demonstrates to the world the obsolescence of the traditional routes of military strength. The very nature of the American forces and their prestigious tools –in particular those which effect a rapid, accurate and crushing defeat, or in other words those which permitted a spectacular, though fleeting, success in the first phase– later proved to be at best inadequate, but more frequently unsuitable or even counter-productive.

Thus the build-up of strength seems to have become pointless, as does the continual growth in military investment. Since the fall of Baghdad on 9 April 2003, the United States has spent 15 million dollars per hour in Iraq! According to some American analysts, between 2004 and 2006, the United States spent as much on fighting IEDs (improvised explosive devices) as it did on the entire Manhattan project, which produced the first two nuclear bombs (used on Hiroshima and Nagasaki), while Al Qaeda, despite all its wealth, would probably be unable to afford a single F22-Raptor (the latest US Air Force attack aircraft). Overall, in the period from March 2003 until now, the Americans have spent more than they spent on the entire Second World War, for a result that is in no way comparable. In that conflict, three powers –Germany, Italy, Japan– were brought to their knees in two and a half years, while the current conflict can at best be described as a strategic dead-end. Even if American defence expenditure is very low in terms of a percentage of GDP[1], the cost of the Vietnam War and Korean War has already been exceeded. The overall budgetary envelope for 2008 is over 600 billion dollars, which is an increase of 10.5% over 2006, and 62% over 2001. War on an industrial scale has become very expensive, even for rich countries. On the other hand, while the industrial resources engaged in asymmetric warfare are very costly, because we are not in control, this war is costing the opponent virtually nothing as he is able to retain the initiative, and to choose the time, the place and the form of the action... He is able to apply the principle of the economy of forces!

The lessons which have been learnt progressively over the last fifteen years of engagements involving Western forces seem to have developed into somewhat of a caricature. This raises the question of the utility of military force[2], as currently used by Western military institutions. In many cases it would see to be inherently counter-productive. Disconnected from its prime role –the political role– by forty years of Cold War, and focused on technical aspects, the effectiveness of force is still based on death and destruction –technical capabilities which have become ends in themselves– which is immediately exploited in the media by the opponent.

Iraq is a textbook example –the worst– but it is not the only one. Once again, after Bosnia, Kosovo and Afghanistan, we see that a brief period of dyssymetric (or quantitatively unequal) warfare, where traditional Western strength has little difficulty in gaining an initial tactical victory, is now transformed into an asymmetric (qualitatively dissimilar) warfare, with a very different kind of opponent. By following the fundamental rule of war, which is to avoid if possible combat with equal opponents, and making a deliberate decision to not to engage in this, the new enemy draws his strength from steering clear of power. He plays with the disparity of resources and of modes of action. Yet again, looking at Iraq, we see a dramatic change in this respect. In the past, the coercion phases were the essential element of an intervention and weapons of destruction were the major constituents of military and political effectiveness. Today, technical and quantitative superiority have been replaced by qualitative and psychological superiority. The

---

1. Between 1942 and 1945, the annual expenditure was equivalent to 30% of GDP, for the Korean War 14% and for the Vietnam War up to 9%, while the cumulative cost of the wars in Iraq and Afghanistan is equivalent to 1% of GDP.
2. An interesting discussion of this subject can be found in the work of General Sir Rupert Smith, The Utility of Force, Penguin Books, London, 2005.

notion of "stocks of strength" has given way to that of "channels of influence". Thus, what we see is that nowadays the role and position of military power are very different from those to which we are accustomed. Military action, so central in the past, has become only one part of the whole picture. The aims associated with initial military success will have to be rethought, as they no longer lead directly to the strategic objective.

Operation Iraqi Freedom demonstrates all the repercussions of the misconception that the brutal application of crushing military strength makes it possible to create easily a peaceful situation more in line with our own interests. The July War in Lebanon, Iraqi Freedom, uneven progress in Afghanistan, problems encountered in crises in Africa... In all these cases it is Western power, with its privileged means of expression and military strength, which is challenged. Rethinking the conditions of military effectiveness becomes essential at a time when we are witnessing a dramatic rise in the cost of modern equipment, with no corresponding increase in its effectiveness. In his book *Irak, les armées du chaos* [Iraq, the Armies of Chaos], Michel Goya gives a clear summary of the situation: "The economist Schumpeter characterised a crisis by the fall in results when using constant resources. According to this criterion, the Western armies engaged in the Middle East are, without any doubt, in crisis. Despite their different approaches, both the Americans and the Europeans have been held in check in Iraq by a few tens of thousands of guerrilla fighters; they have also experienced major problems in Afghanistan. [...] The Western model of war is in a phase of falling productivity, as evidenced by the conflict between Israel and Hezbollah in July 2006. The American paradigm of the remote war, relying heavily on air power, now proves to have a deplorable level of cost-effectiveness. The cost of aircraft and munitions is exorbitant, while the 'targets' have learnt to escape the attacks thanks to a combination of stealth on the ground, physical courage and modern technology." Along the same lines, when describing approaches using means which have tragically lost all political significance, approaches which reject the constraints of a clearly identified enemy and which play no significant role in the analysis, the American commentator Ralph Peters was certainly not putting it too strongly when he wrote in June 2006: "We have become so enamoured of the assets we designed to wage ideal wars, that we no longer see their irrelevance in real conflict."

We are thus witnessing changes in the useful dimensions of power. Once one has moved from wars between states to wars between peoples, or even between ethnic groups, with competition being replaced by challenge, it is no longer a matter of destroying a state's elements of power, but rather of persuading, with the help of force, and helping to re-establish the social contract: winning over rather than submission. Political dialogue no longer relies on confrontation, but instead is established through communication and contact, backed up by a show of force if necessary.

This shift seems to self-exacerbating. The more an opponent deviates from the norm on which force is traditionally based, and the more he diverges from the standards of Western power, the more irrelevant the latter becomes. Nowadays, it is easy to destroy without however vanquishing and to possess superior technology without winning. A defence capability is no longer measured solely in terms of the cost and sophistication of weapons, a state of affairs that changes the impact of new technological advances. Action against new kinds of threat can no longer be restricted simply to eliminating them; there is thus a growing gulf between traditional military strength and the effect one can expect to achieve with it, especially since the increased importance of the local and the tactical itself reduces the utility of the strategic.

In its opposition to this power which threatens to submerge it, social violence finds remarkable sources of vitality. The more it is stifled, the more it flourishes. The more social violence is the consequence of unequal strength, the more it is likely to be strengthened and to move away from the forms that traditional strength is designed to combat. The more such strength confronts the weaker side, the more likely the weaker side is to abandon the conventional rules of confrontation. If, unable to attack the true source of violence, the stronger side uses its ability to deploy conventional weapons in "stand-off" mode to enforce punitive strategies, which give only fleeting satisfaction, it does nothing to resolve the fundamental problem. On the contrary, it is more likely to provoke frustration, a feeling of injustice and renewed violence. The fact that conventional Western-style power is facing similar difficulties in Iraq, Afghanistan and Lebanon –three countries with entirely different profiles, apart from their religion– shows that the difficulties encountered are the rule rather the exception these days. In the face of the new geopolitical and conflictual reality of the sort of war that is likely to occur, the question in everyone's minds is that of the impotence –and thus the inadequacy– of certain weapons and procedures.

While the destructive capability of strength seems unaltered –as it was in Operation Allied Force in Kosovo in 1999 and Iraqi Freedom in 2003– strength has only a limited ability to create a new order. Worse than that, it has damaging effects. External intervention using stand-off destructive capability, although unsuited to dealing with the decentralised use of force, adds fuel to the discussions concerning identity and xenophobia, opposing forms of violence which it is more likely to stir up than to contain. The impatience of Westerners, equipped with weapons intended to produce

rapid technical results, destroys the patiently created balance, without providing any positive solution. In these conditions, the dynamics of the discontinuity and the projection of effects often prove counter-productive; in
the long term they make no contribution to the understanding of the environment and the population or to acting in harmony with local cultures.
These trends have not escaped the notice of the Chief of the French
Defence Staff, who stated in Brest on 25 January 2006: "We have too many
resources generally devoted to what is known as an 'in-depth strike'".

While this rejection of discontinuity is being confirmed at the strategic
level, it is also being strengthened at the tactical level. There is no purpose
served today by "to-ing and fro-ing", relinquishing terrain just captured
without truly taking control of it: having only just withdrawn, the opponent returns to regain the terrain and punish the "turncoats". The only
worthwhile manoeuvre is continuity, where the state advances and establishes a position behind the protection offered by its weapons. In other
words, the "oil stain" technique recommended to us for so long by Lyautey. The orders of the French High Command in Indochina reflected this
policy: "Pacification should be sought, provided the objectives are limited
to the villages which are covered adequately by our military positions, and,
in the eyes of the population, the national administration is as present and
real as were the Viet Minh."[1] Once it has established control and won over
the population, force becomes responsible: the action cannot be reversed
because those who have been won over have crossed the Rubicon, and
thus their future and their lives depend on the persistence of the action.
These requirements[2], which have not changed, have implications for the
size of forces required in future wars.

---

1. Note 92/PAC/CVN of the *Bureau Régional de Liaison pour la Pacification* [Regional
   Pacification Liaison Office] (SHAT-10H 3167).
2. They also correspond to the first of the ten directives issued to his units in mid-2007, by
   General Petraeus, commander of the coalition forces in Iraq, which was "Secure the people where they sleep." This is the primary, long-term mission. Once the zones have been
   made safe, they must be monitored and protected round the clock by the coalition
   forces, until such time as the Iraqi police are capable of taking over. If troop numbers are
   insufficient, it is essential to evaluate the situation in each zone, and then to determine
   priorities so that the zones which have been made safe can later be extended. In contrast,
   an article published in the Herald Tribune on 3 September 2007 – "A short-lived victory
   over the Taliban"– reports on Taliban activities when they retake areas which have for a
   short time been under the control of the coalition forces in Afghanistan: "NATO and
   Afghan army soldiers can push the Taliban out of rural areas, but the Afghan police are
   too weak to hold the territory after they withdraw... one of the things that the insurgents
   do when they enter an area is to hang several local farmers, declaring them spies." Thus
   in these new wars the rule is clear: take control of areas that have been regained and do
   not take areas which you are unable to keep under control in the long term.

FROM STATE VIOLENCE TO SOCIAL VIOLENCE

The decline in the concept of military strength leads to questions as to what it was used to support –the concept of the state–, because, until recently, the two seemed to be very closely linked. Since the Peace of Westphalia, the notion of states and strength has governed international dynamics and the use of force. Without exception, conflicts of identity were confined within state borders, their effects only ever spreading to the immediate periphery, with almost no occurrences of insurgency. The world organisation that emerged after the Second World War reinforced this model. Major crises fitted into the framework of bi-polar conflict, in some cases in spite of themselves, while communications systems did not yet make it possible for minor disputes to acquire a global dimension.

In this context, confrontations of strength were confrontations between states, and war was a means of political communication at the highest level. Its instruments were intended to destroy either the symmetrical resources through which military strength was expressed or the mechanisms governing its organisation and direction. The "hot" conflicts of the 20th century, followed by the Cold War, thus confirmed the role of strength at its most simple and enhanced the military act in its destructive dimensions, casting into the shadows the social attitude to conflicts, which had nevertheless been part of French tradition since the days of colonisation. In doing so they eliminated a crucial dimension from the armed forces: man has been progressively removed from the heart of military effectiveness. He has disappeared from the battlefield and become simply another component of weapon systems, whereas for a long time weapons were one of the tools employed by a soldier in a contact engagement with the other side. The Cold War, driven by technology and industrial interests, eventually placed destruction at the centre of military grammar, conferring on it the status of principal argument for both major and less major strategies.

Thereafter, as the incorrect notion of having reached the end of history receded, the centrifugal forces of unipolarity gradually replaced the attraction of bipolarity, while a number of states –victims of the supranational and the infranational– disintegrated, along with the old, simple condition of working together. Globalisation –seen as interference or even as aggression by certain parts of the world– has led to rejection, through discarding the old model, while at the same time the autonomy of peoples, ethnic groups and religions has grown, in a contagious movement of socio-political fragmentation. The growing faults of integration have exacerbated the injustices –real or perceived– of a world order imposed from outside. In parallel, the powerful party –preferring to settle the dispute

from "outside", that is to say, without integrating– has frequently reinforced his status and strategy of refusing to conform.

What we are witnessing is a breakdown of old social contracts, fuelled
by the lack of integration of individuals or even whole sections of society.
This leads to the revival of ethno-political minorities and a procession of
micro-nationalisms. With the collapse of the old order and the crumbling
of traditional political entities, populations have now become both the
actors and the issue. The ordered geometry of states has been replaced by
the unstable and contested geometry of peoples and ethnic groups, while
the new conflicts, born of old claims which have now resurfaced in the
memory of peoples, have finally broken free from their previous subjuga-
tion. Nowadays states –whose increasing number has inevitably led to
their weakening– no longer have the monopoly of force, or of the use of
force. Other actors have established, in the eyes of international opinion,
the legitimacy of their violent conduct. On the margins –both internal and
external– of today's world, we see a mass of fragmentations and of infec-
tious individualisation which cannot be reached by traditional military
strength.

Traditional forms of socialisation existing alongside alternative forms
encounter confrontation and irritation: these are the response to the ills of
globalisation, falling status, increased urbanisation, frustration and humiliation. Kindled by religious fundamentalism and liberated by the growing
autonomy of its actors, social violence has a free rein. It no longer
responds to state regulatory mechanisms, which are gradually declining in
importance. Rather, it distorts the classic forms of war, leading them
instead into unregulated outbreaks of violence. The geopolitical agenda
becomes blurred: the readily understandable intentions of states are losing
ground to the constant growth of non-state parties, whose interests and
aims are far more difficult to analyse. The weakening of state power opens
up uncontrolled areas, which a multiplicity of rebel competitors try to
exploit. Thus the old interstate conflicts, which have become ever shorter,
are being replaced by increasingly lengthy wars of social breakdown,
where suicide attacks regularly illustrate the level that violence can reach,
fuelled as it is by those who wish to exploit suffering.

However, it would be wrong at this stage to speculate on the emergence of global chaos, which must be contained at all costs. This would
ignore the fact that the chaotic perception of the world is inextricably
linked to the identity of the observer – our identity. The Cold War had
many advantages, not least that of making sense. It enabled us to understand conflicts –albeit wrongly in many cases– even as their number
increased. As Daniel Bournaud put it so aptly: "The sense of chaos cannot

be separated from its period in history... The Cold War produced a sort of intellectual lethargy in that it made it possible to understand international relations as a whole. However, while the Cold War made it possible to rationalise conflicts by explaining them, the universalist idealism of market democracy cannot understand why things do not happen according to dogma. The certainty that the market democracy represents the definitive solution to the problems encountered by human societies prevents the Western world from considering real problems other than through the prism of this new ideology, which owes more to morality than to analysis."[1] The chaotic vision of the world is a refusal to understand that, having left behind the reassuring world of the East-West model, we have simply returned to the truth of the world. This vision is thus primarily the reflection of the faults in our analytical matrices and intellectual responses; it gives rise naturally to the concept of shaping the world, which seeks to reinterpret this to make it more comprehensible to our vision of order, international relations and power. Abstracted from reality, this chaotic vision generates in our minds an opponent whom we refuse to comprehend and whom we prefer to shape according to what we are used to. To adopt this vision is to refuse to understand that, if one closes the parentheses of the 20th century, tomorrow's war is the real war. It is also to search yesterday's world and yesterday's arsenals for the solutions to tomorrow's problems.

RE-EXAMINING STRENGTH

We are witnessing a dual phenomenon: state power is being concentrated at the same time as it is being put into perspective. In a world of "complex interdependence"[2], a state, no matter how powerful it maybe in certain areas –military, diplomacy, economics, finance, technology, agro-food production, education, culture, etc.– may prove to be quite vulnerable in others. This relative vulnerability has the effect of reducing the overall capacity for action, and thus overall strength. It seems that today, true strength is dependent on the absence of vulnerability. Yesterday Achilles was slain by Paris's arrow; today we see clearly that the colossus has feet of clay.

---

1. "*Le chaos, un mythe dangereux*" [Chaos, a dangerous myth], in *Agir*, general strategy review, September 2007.
2. Keohane R.O. & Nye J.S., 1997, "Interdependence in World Politics", In Crane G.T. & Amawi A., The Theoretical Evolution of International Political Economy: a reader, New York, Oxford University Press.

In the military field, as long as strength depended clearly on wealth and on the force of weapons, it remained in the hands of the state. Information, which itself has become power, has reduced the dissymmetry between those who traditionally held it –the states– and the other actors. In these circumstances, the concept of power or strength, is no longer synonymous with military power or strength, as used to be the case. Traditional power, seen in terms of military capability or the ability to constrain and deter, is increasingly losing its value. In order to impose its will on the other party, the state must nowadays make use of a wide range of resources, with economic, diplomatic, cultural and social aspects all having a role to play. In addition to this expansion, the concept of power seems to be becoming separated from that of the state, and becoming something other than an ability to impose one's will. It is also the ability to influence, to escape from the will of the other party or even to disrupt, and thus to reduce, the other's means of expression. To be powerful always means having the ability to constrain, but the means to do so have become far more diversified and are easily accessible to the new actors in the political arena. We now know that it is no longer necessary to be a state to defeat a state.

It is true that in the shrinking world of states which accept the rule, traditional power has not disappeared; however, its ability to constrain is dwindling, and its relevance is diminishing dramatically in the grey areas of the new conflict scenarios. Henceforth a state, even a weak one according to traditional criteria, can make use of a myriad of levers to offer effective opposition to a much stronger party, or even oppose the whole of international society. Strength is now measured much more in terms of the strength of "effects" rather than of "resources", the two aspects now being only very loosely correlated. There is little point here in dwelling on the strength of major international companies, but we cannot afford to ignore the terrorist groups, organised diasporas, mafia-style and other criminal groups or small countries whose possession –or pseudo-possession– of nuclear weapons gives them a disproportionate weight on the international scene. Skilled opponents, operating below the horizon of conventional high technology, can thus easily circumvent such technology and challenge a domination they would be unable to confront directly. A large number of factors play a part in levelling out strength and in downgrading conventional military strength to the level of just one of the elements of the much reduced effects obtained when such strength is used against an opponent who refuses to accept the notion.

Conventional strength remains an important trump card, particularly since it offers a range of differentiated options. However, it will only remain an asset if we are able to modify the spectrum devised in the past in line with the changes in the conflict environment; in other words, if we demonstrate situational awareness and an ability to adapt.

Today, in Lebanon and Afghanistan, in Iraq or elsewhere, military force alone –that is, traditional military strength as shaped by the century of total war and the bad habits which set in during the Cold War– is not enough to resolve crises. Given the many changes, the temptation to affect ignorance is great and actually quite reassuring. Indeed, the investments already made, the current industrial dynamics, the well-established patterns of thought and institutional interests tend in this direction. Yet it would have serious implications for the future if we were to misjudge world developments and their direct consequences for the models of force and the conditions in which they are deployed. These days, recourse to strength alone can prove to be a rejection of the diagnosis and a dangerous denial of the complexity of a world temporarily hidden behind the mask of bipolarity.

Naturally the owners and operators of high-tech weapons, from whichever country, see the world in their own way. The more skilled one is in a certain area, the more easily one is persuaded that it is right to follow a particular path. In addition, those states with the greatest traditional power remain conservative in their perception of power and their way of directing it, assuming –or giving the impression– that the potential opponent resembles the opponent of the past. For understandable reasons their instinctive reaction is to leave their military establishments unchanged and to maintain the balance achieved in the past, particularly since the difficulty involved in coming to grips with new threats –more abstract and more qualitative in nature– makes them reluctant to abandon the reassuring certainties of qualitative notions. Armies where technology plays a greater role than manpower naturally see events and methods through this prism. They seek solutions within their own range of capabilities. For those in possession of weapons designed primarily for a war that is unlikely to happen, it is very tempting to pretend that it will.

However, there is a serious risk attached to thinking that "major war" is the only true form of warfare, that the enemy will probably remain stuck in his industrial perception and that his will and creativity can be seen as negligible. As General Jean-Marie Faugère puts it so well: "Being able to limit collateral damage would not be enough to justify the use of weapons which are becoming increasingly destructive. [...] Owning 'prestige' weapon systems, supposedly giving the owner the status of a great power, implicitly incites those in charge to overestimate the reasons for intervening in a crisis, but also to underestimate the repercussions of the military action chosen. Such action is not neutral, neither for international public opinion, nor for the populations immediately affected by the crisis. It determines not only the conduct of the operation itself, but, even more strongly, the exit routes from the crisis, depending on whether these seem to be appropriate for the situation, disproportionate or even downright excessive, given the effects on the terrain and their consequences for the population."[1] One thing is certain, it would be better if the "fog of war", which no technology will ever eradicate, were not made worse by a persistent refusal to see clearly.

It can be very tempting to ignore the real pathology and to stick with familiar therapies in order to be able to make use of the remedies available in large quantities. However, military action cannot be the "continuation of the absence of politics by any other means", restricted to a demonstration of what high-tech forces can do, because this is the technology that gives us superiority. It is clear that tomorrow's war will no longer be a technical confrontation between two arsenals. There will no longer be the close correlation between the strength of the weapons a force has available and the ability to gain the advantage, that is to say, to produce political effectiveness. Where it is not possible to "understand fully the type of war being fought", to use Clausewitz's words, an accumulation of technical strength may be simply an accumulation of political weakness. In the new environment the greatest tactical victories may lead to strategic chaos, or even defeat.

Thus the only solution today is to understand tomorrow's war and to examine the conditions needed for military effectiveness. The high precision, high destruction war on a large scale is no longer an essential argument for effectiveness, since it no longer makes it possible to "conduct politics by other means". Clearly, we continue –in fact increasingly so– to need force, but it must be useful military force. For a state to be capable of implementing its primary security mission, it is essential to ensure the utility of the armed forces and their military systems. This will require us to ask a number of questions and to give clear and bold responses. We cannot hope to resolve tomorrow's problems using only yesterday's solutions. We will not be able to produce an effective strategy or policy for our future wars using only the weapons of yesteryear.

If we truly wish to prepare ourselves for new types of war –those we are likely to encounter and those we will have to fight, rather the ones we dream of because we know how to fight them– we will need to make dramatic changes in our way of thinking, so that we are not answering questions that are no longer being asked. We will have to change our thinking.

---

1. In *Inflexions*, January – May 2007, *La Documentation française* [French documentation], p. 185.

CHAPTER II

NEW WARS

What wars and crises will the 21st century bring? This subject has long been exercising thinkers, strategists, geo-politicians, prophets and other forecasters. My intention here is not to put forward a new and original response to what happens when potential crisis situations arise, but rather to go beyond ideas focussed solely on the risks linked to terrorism –which is only ever a mode of communication, albeit a horrific one– and proliferation.

Given that we are unable to determine where tomorrow's war will break out, an awareness of its general context will provide some indication of its likely style –the only element we are reasonably able to anticipate, since it is virtually impossible to determine the circumstances of future crises– and thus of the changes we should make to our forces if they are to regain full utility.

GROWING IMBALANCES LIKELY TO LEAD TO WAR

Examining the root causes reveals that, as a result of the new dynamics among populations, demography –for so long the origin of all crises and wars– has returned to the top of the list. With the world population likely to rise from 6.5 to 8 billion within the next twenty years, the demographic factor is now one of the main elements determining geostrategic developments. Further, migration continues to increase as a result of the exacerbation of inequities and the accompanying political, social and economic crises. The rapid growth in urbanisation –in thirty years time over 65% of the world's inhabitants will live in urban areas– will lead to other internal imbalances, which could well create considerable tension with regard to the distribution of resources. Potential conflict situations will arise from social, ethnic or economic fragmentation, and malnutrition and epidemics will lead to humanitarian disasters. In addition to crises caused by the movement of people, there will be –on occasion both local and general–

difficulties relating to population size, since the ability of the planet to feed and support us is finite... Thus there will be problems with both movements and stocks!

Increasingly, scarcity of resources, combined with growing demand, will become a source of potential competition and conflict. Energy resources –in particular fossil fuels– represent a strategic asset, made even more fundamental by the fact that emerging countries are consumers rather than producers. Thus competition for access to oil and gas fields and for control of supply routes will become tougher. The unequal distribution of food resources will get worse over the coming decade and, in the international context, will be a further important factor likely to spark off a crisis, made even worse by the risks to health. In parallel, while international development aid is stagnating, or even declining, developing countries are seeing a fall in education spending, thereby increasing the risk of extremist movements exploiting uneducated populations.

Furthermore, the climate factor also has an accelerating and exacerbating effect. The trend towards increased warming, coupled with human practices that are destroying the global environment, are seen as damaging for general security. The ensuing disruption –desertification, flooding, scarcity of arable or habitable land– leads to greater inequities through increased population movements, this time of "climate refugees". Here too, there will be violent competition for the ownership of essential goods. It is true to such an extent that combating climate developments acts as a direct means of crisis prevention. These are likely to become the most serious security problem since the Cold War. On 23 October 2007, the German Minister of Foreign Affairs, Frank-Walter Steinmeier, put the point forcefully: "Policies to fight climate change can, and will, become an important part of peace policies." [1]

Together, demography, the fight for resources and the climate factor are responsible for increasing imbalances, or in other words, growing inequality. It is these imbalances –between populations and resources or between populations and powers– in both inter-state and intra-state contexts, which give rise to conflicts intended to regulate or equalise matters between states or within states. Such imbalances are naturally aggravated or politically exploited by fundamentalism and demands for recognition.

Thus the cumulative effect has moved us from crises with a primarily political background to systemic crises loaded with political repercussions. They are superimposed and intertwined in a complex architecture, in which no crisis is resolved before the next one makes its appearance.

----

1. Herald Tribune, 23 October 2007.

Somewhat oddly, these effects are worsened by the effects of modernity. The growing trend for "transparency", which has been made possible by progress in the field of communications, renders that which was previously little known clearly visible to all. Acting as a catalyst, it raises expectations more rapidly than one can respond to them. Today, frustrations previously blurred by ignorance are shamelessly exploited. More and more clearly, we see that globalisation is primarily benefiting its initiators, usually at the expense of those who are excluded. Humanity, largely abandoned at the cost of progress, can only observe with a growing sense of injustice and expectation; it sees that, for the most part, it is in danger of being permanently excluded, while at the same time suffering from the deteriorating environment, which is one of the consequences of globalisation. So, in a world where opinions count for more than facts, the expansion of the "egalitarian passion"[1] carried by the emotions roused by the media leads each day to fresh frustration and violence, with the gap widening at the same pace as the means of comparison become simpler.

In the West, we live in an artificial security bubble, well protected from the outside world. This allows us to think that tomorrow's war is something that will concern others; victims of their own success, defence and the efforts required would be pointless. But we should indeed be worried about the future. The inequalities that have grown up as a result of the highly unequal distribution of growth between regions, on the one hand, and between the social strata in various states on the other, combined with the new evidence of disparity, will lead to tension and confrontation in the future, both elsewhere in the world and on Europe's doorstep. The frustrations felt by those who gain no benefit from globalisation will be directed against governments, while impoverished states will face the expansion of nationalist and religious streams taking advantage of the collapse and failing legitimacy of state structures.

If we take no interest in this situation, the crisis factors will get worse; although many people do not realise it, we are already fighting tomorrow's war, in a number of theatres.

Hot spots close to France

French national security will continue to depend closely on the security situation on Europe's doorstep, while at the same time its Mediterranean location will render it vulnerable to crises and conflicts in the Middle East, with ramifications extending far beyond the immediate area.

---

1. Delpech, p. 46.

Because of the porosity of Europe's borders, the Balkans must be seen as our next-door neighbour. The crises destabilising that area present us with direct security concerns. In particular, whatever the future of this area may be in Europe, France will have to be able to contribute, with military means if necessary, to establishing the stability essential for its own peace of mind.

Although the Maghreb area is geographically and historically separated from the rest of the Arab world, and thus marginalised with regard to the major strategic stakes at play in the Middle East, it nevertheless faces the same challenges as the rest of the Arab world, principally in terms of political openness and social and economic development. Their effects on regional stability will directly affect Mediterranean Europe.

The American intervention in Iraq exacerbated the latent instability in the Middle East and paved the way for transformations with implications that cannot yet be fully grasped; all that we know is that there will certainly be serious consequences for France. Iran's current attitude is of course linked to its perception of the strategic impasse in which the United States currently finds itself, given its presumably long-term inability to engage in a further conflict, other than with stand-off strikes. In the long term, such strikes prove unproductive, as demonstrated by the US interventions in Sudan and Afghanistan in 1998. Despite this, France will continue to need a sufficiently stable Middle East, notably for reasons of energy supply, and may find itself required to be part of a military coalition acting to achieve this.

In spite of the scale of its natural resources, Africa –the continent which is likely to see the greatest population growth– will for some considerable time remain a highly unstable continent. Demographic inequalities, refugees fleeing from armed conflicts (most often intra-state conflicts), the weakness of states and institutions, various forms of trafficking, and the growth of diseases and pandemics associated with poorly controlled demography all contribute to mounting sources of potential conflict in the medium and long term. France cannot wash its hands of a continent which, despite some democratic and security progress, will require careful guidance, particularly in military matters, if it is eventually to be able to take charge of its own security problems. Aside from its energy potential, which France may wish to take greater advantage of in the future, the various French communities across the continent make it essential for us to retain the means to protect, and if necessary to evacuate, our nationals.

Not understanding the new forms of war means not being able to prepare the correct tools, and thus losing future engagements. Defeats are rarely solely technical – they are primarily the result of a lack of understanding. History, both recent and more ancient, should act as a warning, since it indicates that we are unlikely to be naturally skilled at fighting the war of the future. We should also note that, albeit in a totally different political context, Western counter-insurgency forces –notwithstanding their military and technical advantages, and sometimes their brutality– have almost always failed in their endeavours. The one notable exception is the British forces in Malaya during the 1950s. Nevertheless, we cannot accept the abstract notion of reducing our engagements. Despite these problems and the difficult legacy we face, and even though there is no way of knowing in advance where, when and for how long we will be called upon to intervene, we must remember that the fate of the citadel is to be surrounded and then destroyed. The simple defence of staying close to home is doomed to failure. What would be the reaction to a football team relying solely for its defence on the goalkeeper and a couple of defenders glued to the goalposts? In other words, according to one of the key strategic principles, one should defend as far forward as possible and always "in the deep area", on the "outer circles", while being prepared if necessary to defend while withdrawing to the "intermediate circles" before, if required, putting up a "firm defence" of the innermost circle, that of the protection of national territory. Defence Minister Hervé Morin clearly took the need to go and defend "from the front" seriously, when he said: "We need to contribute to stability beyond our borders."[1]

This forward defence will force us to intervene in future conflicts, in distant arenas where we will be required to contain and, if possible, to reduce the threat. It will take us to intermediate spaces and maybe also to our borders. We can see on a daily basis that confrontation and war are an integral part of human nature. All the signs indicate that the potential for crises is bound to increase. We are moving towards a world which is less stable, with a greater number of possible wars, thus justifying the effort put into ensuring a "useful" defence capability. This vision is share almost unanimously. General George Casey, Chief of Staff of the US Army, speaking at the National Press Club in Washington on 14 August 2007 was particularly enlightening on this topic: "I sent my expert officers to talk to specialists in universities and think tanks. They went to intelligence agencies and to the government. They came back surprised at the near unanimity that the next decades we face here will be ones of what they call persistent conflict."

Since there are currently no forces which correspond to the type of conflict we are likely to encounter, restoring the utility of the armed forces

---

1. 11 September 2007, Toulouse, Defence summer school.

means adapting them to the engagements they can expect to face. Naturally such forces must retain an ability to increase their strength and to react to conflicts which are unlikely, but not inconceivable. But what would be the point of excelling in a type of war that just might occur in the next twenty-five years, if we are defeated long before because of our inability to respond to the immediate challenges? We must therefore understand the characteristics of warfare tomorrow. General Jean-Louis Georgelin, Chief of the French Defence Staff, takes a more pessimistic view, envisaging a future of long and difficult conflicts: "We are likely to find ourselves taking part in complex conflicts, where the stakes are crucial to our own security and which will involve prolonged contact with the enemy. These new conflicts could thus combine the intensity of peace-enforcing operations and the length of peace-keeping operations."[1] Thus we have to prepare ourselves for tomorrow's war, the circumstances and details of which we have no hope of being able to predict. The gravest error we could commit would be to give in to the desire for simplification in an attempt to determine our inevitable success on the basis of an excessively theoretical greatest common denominator.

A symmetrical war is unlikely

Looking at the characteristics of the type of war we are likely to have to fight, the first key idea is that the traditional warfare –as we understand it today, with one industrial power facing another– for which we have honed our weaponry and our forces is probably over.[2] The symmetrical war is dead, or at least the chances of it happening are negligible. The Chief of the French Defence Staff put it clearly when speaking to the cadets of the naval Academy on 25 January 2006: "War, or rather a declared war involving belligerent populations engaged en masse, nation against nation, is, in my opinion, history." Taking this further, one could say that perfecting warfare has killed off the traditional war, fought using equal weapons. The stalemate reached on 9 November 1989, with the collapse of the Berlin Wall, was followed by the checkmate on 9 April 2003, with the fall of Baghdad. There are two kinds of reasons for this: economic and nuclear.

In the first case, we have already seen that traditional war is no longer a cost-effective means of realising political or economic aims. In parallel, the meagre return on each additional dollar or euro invested in defence is

---

1. *Politique internationale* [International policy], no. 116, Summer 2007.
2. In spite of their usual differences of opinion, it is interesting to note that British (DCDC) and French (DAS/EMA) official thinking concurs on this idea. See footnote below.

diminishing. Indeed, on the one hand, the exorbitant cost of modern warfare renders it less and less efficient, at least on a large scale. This was discussed extensively in Chapter I. On the other hand, remaining in the area of economics and using its language, globalisation reduces the likelihood of a large-scale war because it increases the cost of such a war, through a lasting decline in trade and its ensuing benefits. Moreover, regional integration increases economic interdependence and, in so doing, also reduces the probability of conflict between neighbouring countries. Europe offers the ideal model, where the interdependence sought has created peaceable coexistence. Even more profoundly, in an increasingly interconnected global system, the considerable damage that a conventional war would cause to both winner and loser would have intolerable economic consequences, out of all proportion to the desired results. Thus, with regard to the "major war", globalisation now more or less fulfils the role previously played by deterrence. British official thinking has adopted this idea. In the 2007 edition of "Strategic Trends"[1], the DCDC –a Defence Staff think tank– says that: "Major interstate wars will be unlikely (in the next thirty years), because of the increasing economic interdependence of states in a globalized economy." At the same time, a similar French document, a joint DAS/EMA paper[2], adopting a cautious style intended not to offend conservative thinking, states in June 2007 that: "An exhaustive study conducted for each country does not necessarily lead us to envisage any situation which could degenerate into a conflict on a major, far less world, scale."

For the second category, it is probable that, at the level of the so-called "macro nations", nuclear deterrence will continue to play its fundamental role of reducing the potential level of conflict. If, conceivably, we were to hold an opinion to the contrary, it would be wise to rid ourselves immediately of the very costly nuclear arsenals we maintain at the expense –literally– of our conventional capacities. It is worth pointing out, however, that in the past, the deterrent acted against a lethal threat and, although it still does[3], this lethal threat is no longer the main threat. The United States has have been acting on this since 2002. The US "Nuclear Posture Review"[4] sees nuclear deterrence as one element of a new triad of "enhanced pro-

---

1. The DCDC Global Strategic Trends Programme – 2007/2036, DCDC, January 2007, p. 68.
2. *Préparer les engagements de demain – 2035* [Preparing engagements from tomorrow to 2035], a joint document issued by the *DAS* (Strategic Affairs Delegation) and *EMA* (French Defence Staff), June 2007, p. 46.
3. There is thus no alternative to maintaining and reinforcing the deterrent effectiveness of our nuclear arsenals, though at a level of minimum sufficiency.
4. Nuclear Posture Review, 8 January 2002.

tection". Since nuclear deterrence deters the lethal, but not the most likely, threat, a complementary deterrent must be put in place. This can be done by constructing conventional forces capable of responding to the challenges of tomorrow's war, and thus of deterring the likely opponent.

For this reason, as well as for simple reasons relating to the utility of the forces and the investment they represent, the decreasing likelihood of encountering a "conventional" war should lead to a certain amount of readjustment within the forces. Indeed, even though the land forces have suffered more than other services from the disastrous consequences of the flawed notion of the "peace dividend", we see that since the end of the Cold War all armed forces have essentially undergone a corresponding reduction. This distortion has caused a growing discrepancy between the model and the reality of crises, which have certainly not followed this pattern.

We note, however, and this is fundamental, that putting the traditional war "on ice" implies that, for some considerable time to come, we will need to retain the capacity to fight such a war. It is difficult to "un-invent" modes of war. Just as we have been unable to un-invent nuclear war, and the presence of nuclear arsenals strangely continues to justify their existence, we cannot un-invent traditional war. Thus, only the continued capacity to conduct one, and to conduct it well, will reduce the risk of us having to do so. Just as war does not abandon a space (land, sea, air, information, etc.) once it has taken it, a second deterrent layer –in this case a conventional one– is deemed to have been added to the first. And in the same way that it was necessary to invest in nuclear and conventional weapons, it will be necessary in the future to invest in forces suited to these conflicts –the possible conflicts– while at the same time constructing forces capable of demonstrating political effectiveness in the probable wars, those we are likely to conduct. It is not a case of one or the other, but rather of both, in proportions which remain to be determined. This is all the more true since, while the war we are likely to have to fight will certainly not be a major head-on battle, it is bound to include many high-intensity engagements.

It is even more true, that to reject once and for all the possibility of a major war, would be to ignore both the history of mankind and the very nature of war. The peripheral fragmentation of fragile states does not mean the death of the old states and, outside the Western environment, military rivalry between states persists. The possible resurgence of malevolent states equipped with key elements of conventional military strength could combine with Europe's congenital energy dependency to form a direct threat to our strategic interests or our strength. Thus the possibility of a major interstate conflict cannot be completely excluded, even if the cost of such a major war has become rationally unacceptable among developed societies. Furthermore, this event remains dependent on conditions which bear no relation to historical need. We know that war has often been the result of an uncontrolled sequence of random happenings. We also know that war can easily take on "a life of its own", enabling it to escape the limits within which we had hoped to contain it. War by no means obeys the simple rules of rationality; its very unpredictability means that we must be cautious when attempting to predict the future.[1] Thus today, we should not get rid of the means to fight this type of war, and we must ensure that for the future we retain the capacity to build up strength, particularly since we cannot exclude the possible re-emergence of a bi-polar rivalry at some point during the next half-century. Deterrence and constraint therefore remain two fundamental aspects of any defence system. This point of view requires us to have the latest equipment and to maintain our lead in research and development, through a constant, judicious command of technological progress. Should it become necessary to constrain or essential to destroy, the increased need to regulate the effects, by means of accurate, rapid and proportionate action, obliges us to have access to the most effective means of imposing our will.

It would, however, be wrong to assume that a future confrontation between two strong parties would necessarily resemble the sort of confrontation we prepared for in the past. To do so would be to reject the basic rule of avoidance, as we unfortunately did in the period between the two World Wars. Intellectually speaking, superiority is a dangerous thing. This is one of the things that the Israelis were forced to re-learn in 1973, when they almost lost the Yom Kippur War against Egypt and Syria, whom they had roundly defeated six years earlier. In other words, it is important to be fully prepared for the possibility of a "strategic surprise", and this preparation should include a good understanding of what such a surprise might entail. The war is unlikely to follow the patterns of past wars –or it would not be a surprise– and thus we will not be able to use the models we had prepared for use in such circumstances. It is also likely that this "surprise" will not be a conventional war between two strong parties, otherwise this would imply acknowledging the futility of our deterrent. It is also likely that, even in the event of the re-emergence of a major military threat, the strategic surprise could not be a symmetrical war in which we could use our strength to its full effect. Indeed, any opponent who opted to confront us as an equal would only do so if he were certain of being able

---

1. The DAS/EMA document referred to earlier states (on p. 46) that, "it is impossible to exclude the possibility that an unfortunate combination of circumstances or set of alliances may drag France and the European Union into a major conflict."

to reduce our superior strength immediately –in particular our Achilles' heel, our networks– through information warfare or weapons using electro-magnetic pulses. Thus preparing for the strategic surprise means, firstly, recognising that our opponent will be intelligent. He will choose, where possible, to operate in areas which are "off limits", focusing more on movements than stocks. In addition, if he opts for a military engagement, he will strike at our weak points directly "levelling out" or neutralising our "comparative advantages", and avoiding our arsenals and our main strength by following the age-old and fundamental rule of war: circumvention.

From symmetry to asymmetry

In the face of an possible opponent who is likely to reject the verdict of the battlefield, strength and power are always complex. Thus, although traditional military strength, if not supplemented, can be bypassed, it remains indispensable. To ensure their utility, the new force models have no choice but to include both "intelligent weapons" and significant numbers of strong units.

Strength does indeed seem subject to a major paradox: logically, it should be continually reinforced, but to do so alters its relevance, with an excess of strength leading to practices which aim to challenge it and then to circumvent it. When faced with a power too strong to risk attacking it with his own weapons, the weaker party, considering the stronger to be an unsuitable target, invents new forms of challenges, some of which alter the concept of victory. He delays the decisive action, avoids defeat, expands and reinforces his network and establishes his internal and external bases in order to turn the overall balance of force in his favour. Destructive power, with its imbalance and its perfection, provokes avoidance which, in return, renders it pointless. This phenomenon is by no means new. Clausewitz told us that when choosing a battle position one should: "Select a battlefield offering us advantages; but for this position to become a battlefield, these advantages should not be excessive. [When on the defensive] by keeping our armed forces in an impregnable position we simply refuse to join battle and thereby oblige our opponent to seek a decision by other means."[1] In the same way, the opponents in tomorrow's wars will naturally refuse to fight the high-tech wars that we are preparing. Everywhere, the irregular opponent refuses to subject himself to the exceptional destructive power of our modern military forces, causing the perfect power to fall into the vacuum created by the opponent. Rather than engag-

---

1. Quoted by Durieux, p. 39.

ing in high-speed, brief confrontation dominated by Western forces, the opponent we are likely to encounter prefers to invest in long political struggles and psychological trials of strength. He plans over decades and avoids our battle as he has no need of tactical victories.

The growing dissymmetry between the arsenals means that the weaker party has no hope of winning. For the irregular opponent, the impossibility of responding in the same register creates a sense of injustice and serves to exacerbate the violence. Stand-off action, well beyond the reach of any return strikes is seen as an act of cowardice, legitimising, in the eyes of the perpetrators, actions in response which do not comply with the Western law of war. Thus, the more widely strength is deployed, the more it mobilises feelings against it, provoking resentment, dispute and circumvention, rather than confrontation. In contrast to the logic that prevailed in the last century, "symmetric power" is perceived as "asymmetric", rather than "dissymmetric" by those subjected to it. It provokes "asymmetric" reactions –which do not need the violence delivered by high technology– rather than a search for "symmetry", since the desire to be at least equal, which drove Soviet psychology, no longer exists. Given Western technological superiority in the four traditional operational areas (land, sea, air, electromagnetic), the opponent has no choice but to develop strategies enabling him to bypass these and to find areas where he can fight on an equal footing. The "infosphere" and the human area, or space, are where the future war will develop. The historic experience encapsulated in Clausewitz's law of reciprocal action tells us that it is not possible to bring the opponent into our own territory, so if we wish to be rid of him, we will have to seek him out in his own territory.

This reasoning led General Hagee, Commander of the US Marine Corps, to say in 2005, "we know that our future will be dominated by unconventional wars". Secretary for Defense Gates, in a speech on 10 October 2007[1], confirmed quite clearly that: "Unconventional wars will be the ones most likely to be fought in the years ahead [...] We can expect that asymmetric warfare will remain the mainstay of the contemporary battlefield for some time. These conflicts will be fundamentally political in nature, and require the application of all elements of national power. Success will be less a matter of imposing one's will and more a function of shaping behavior – of friends, adversaries, and most importantly, the people in between."

Thus, the tools and methods we need to devise and have available if we want to intervene in the world and impose a political way of thinking

---

1. AUSA fall symposium, Washington DC.

are different from the ones we needed for past wars. Tomorrow's war –the war we are likely to have to fight– will take place on the ground and will mean close combat since, despite all the technological miracles, to defeat a man armed only with a dagger, one must draw one's own. Modern Western societies had only an abstract and virtual link with the violence of war, but in tomorrow's war our adversary will have a face; it could be the face of hate, because we will be his "sworn enemy" and he will consider his and our continued mutual existence as incompatible. We must understand that asymmetric warfare is not some degenerate form of war; it is simply war and, looking back into history, we see that, over the ages it has been the most common type of war.[1] This apparent return of the asymmetric war is actually just the end of a process of concentration. Since 1945, conflicts which we have referred to as "low intensity conflicts", treating them with a certain degree of disdain as we did not consider them true wars, have not only been the most numerous, but also far and away the most deadly.

It is interesting here to draw a parallel, though some may consider it a little rash. The victories achieved by the French Republic and the French Empire can initially be explained as a clash between two models, where one proved to be superior: that of the small, princely armies –professional and costly – versus armed nations– popular armies, simple but numerous, born of a national ideal. This clash of paradigms –Westphalian wars versus popular wars– is being replayed today, but in reverse. It is well-known that it was not until the shock of the defeat at Jena (1806) that the princely armies realised the need to change and adapt to the new models. Having thus adapted to the developments of the other side, they eventually managed to deal with the brilliant troublemaker, finally defeating him nine years later, using the concepts he had earlier used to his own great advantage. Thinking about this allows us to draw hope from our intelligence and from the future.

It is quite true that, even though traditional war is probably dead, in the initial phases of a conflict, Western forces will still need to conduct dissymmetric engagements, in which they will face forces of a similar type, but

---

1. Despite its natural penchant for "major war", American official thinking has now also accepted this point of view. The official document, "Marine Corps Operating Concepts for a Changing Security Environment" (Quantico, Virginia, 03/06) considers that: "Conflicts like World War II represent both an aberration as well as a refinement of the actual tradition of war. The traditional form of war is actually more irregular. [...] Throughout American history the default setting for military preparedness has derived from what was considered conventional or regular." It goes on to say: "Future conflict will not be dominated by tests of strength that characterize industrial war. It will be dominated by wars fought among the people... Irregular threats will likely be the predominant threat we will face in the future... Irregular wars will characterize the foreseeable future."

with much more limited capacities. On the other hand, what we note is that, without exception, dissymmetric engagements, which are always short, are now turning into asymmetric conflicts. In these often long and costly conflicts, the conventional force faces an irregular opponent using heterodox modes of war. There are a number of recent examples, such as the war in Kosovo: two months of conventional war and, to date, eight years of asymmetric crisis. In Afghanistan we have had one month of traditional war, followed by, so far, six years of asymmetric crisis. The war in Iraq has been three weeks of traditional war, followed by five years of asymmetric war.

Today, we are observing a significant change in the proportions. Yesterday the coercion phases were the key element of an intervention, since the aim was to constrain a state, and to destroy its military capability in order to achieve this; the means of destruction were thus the mainstay of political and military effectiveness. Today, in most cases and for most of the time we are not facing this type of opponent; instead we are acting to restore a state, on behalf of its population. The initial high-tech battle is short, but the campaign that follows it is long, even if it is also high-tech.

What we also know is that in the conflicts we are most likely to see in the future, the role and the place of military action will be very different from we had become accustomed to. Military action, which in the past was key, is now only part of a whole, or in military jargon, one of many "lines of operation". It still remains essential, because the state of a country's security determines reconstruction efforts, but is now only one of many lines of operation. Others include diplomacy, economics and humanitarian aid. In the past, when classical philosophy perceived war from the point of view of the state, the role of a military victory was to lead to a political victory; in the future it will be the reverse. We therefore need to look at things differently.

Our opponent determines our laws

What has been described above are general trends. It would be a very grave error to draw from these any assurances as to the exact nature of either the war or of our likely opponent. Donald Rumsfeld, the former US Defense Secretary, on taking up his post in January 2001, already felt that: "In an increasingly unpredictable world, we find ourselves in an era of the unexpected and the unpredictable."

Uncertainty is not, however, the only certainty: we also know that two rules apply invariably. The first is that of the intelligence of our opponent. We should show intellectual respect for our future opponent: respect for his capacity to reflect, deduce and adapt, as well as for his ability to identify our strengths and our weaknesses. The second is that of circumvention. The very principle of success in war is that of avoidance because, while for tactical success it may be enough to destroy the enemy's force, strategic success is usually achieved by bypassing it – by means of the second dimension, the third dimension, through the economy, technology, or by some other means, but almost always circumventing it in some way. Circumvention in space in two dimensions is illustrated by the success of Hannibal in Cannes or General Schwarzkopf in the Iraqi desert in 1991. It was also the idea behind the Schlieffen plan, which nevertheless failed due to the inadequacy of the resources available to implement this grand vision. Circumvention in the third dimension is in line with the philosophy of Giulio Douhet, the Blitzkrieg in 1940 and even Hiroshima. Circumvention is also possible in the other areas that man can control, for example, circumvention by means of technology. During the Cold War this was at the very heart of the race for superiority between the two super-powers, and nowadays it is illustrated by the investment in "new" bio and nano technologies, to name but two areas. Circumvention is also possible in the outer atmosphere, in the infosphere and in the electromagnetic spectrum. In this last case, the remarkable effectiveness of cyberattacks, and the speed at which they are increasing, should lead us to consider the vulnerability of what we see as our source of strength: communication networks and new communication technologies. Circumvention, notably of the "rules" which we have adopted uncritically; the first duty of an intelligent opponent –and we can be sure that they all fit into this category– is to be "irregular". Without any doubt, circumvention constitutes the very nature of an "off limits" war, the sort of war we are likely to have to face.

More than any other, this war of "mutual circumvention" will place great demands on the principal characteristic of the soldier: adaptability. It may make it possible to avoid the scathing criticism heaped on the imperial troops fighting Spanish insurgents during the winter of 1808-1809 by Baron Jomini: "While, like Don Quixote, you are tilting at windmills, your adversary is heading for your line of communications, destroying the detachments left to guard it and surprising your convoys." Thus, while past wars –technical wars of destruction– were technically complicated, tomorrow's war is likely to be complex, requiring increased adaptability. The particular circumstances of a war have little effect on the destruction machine –an aircraft does not change according to its target– but they have a continuous effect on people and their interaction with the conflict, in both camps.

Such circumvention is simply the eternal truth of war: in essence every victory is asymmetrical. The supposed irregularity of the opponent is only the application of the sole lasting rule of war, at least of the probable future war, rather than of the one we would prefer because we are more familiar with it. War is most definitely not a sporting encounter conducted according to rules accepted by both sides: there is no referee to blow a whistle to indicate that the end of the mutual circumvention session has been reached. There is no first or second place, but rather a winner and a loser. The logic of war, understood so perfectly by Clausewitz, differs from political logic; it is the logic of the duel where "each adversary observes his opponent's actions". It is therefore impossible to define in advance or unilaterally its space, its range or the type of violence that will be deployed. Internal dynamics have a greater influence on the conduct of events than political will. Tomorrow's warfare will be "irregular", not only because it will involve "irregular forces", but because it will conform to no "regular" pattern.

This rule of circumvention reflects precisely the spirit of the excellent title given by two Chinese colonels –Qiao Liang and Wang Xiangsui– to their visionary work: Unrestricted Warfare. The strategy described consists, for the weaker side, of investing only minimum effort in the military sphere, where he knows he cannot win, and, in keeping with the times, of concentrating his actions in other areas where the stronger side is vulnerable: information, economy and finance, politics, etc. Let there be no doubt, future wars will be "unrestricted". Anyone thinking there will be a head-on clash against an equal adversary –as in the wars of the past– has no real insight into the phenomenon of the warrior or human intelligence. Even taking only the example of China –which has now become both the banker and the supermarket of the West– we see that it has many ways of achieving the aims ascribed to it, making it unnecessary to seek confrontation in areas where it is fully aware that its chances of success are at best negligible. Let us be under no illusion – there are few adversaries who would be happy to become the consenting victim of the sort of war at which we are so skilled, which we would like to conduct and for which, in many cases, we are still preparing.

Tomorrow's wars to reconstruct the state

While symmetric war is unlikely, we have little chance of completely escaping a future war: the end of the reign of the major inter-state war combines with the absence of peace, its effect, to multiply the number of "states of violence"[1] in the world. For this reason, attempting to avoid the likely war at all costs amounts to inviting it into our home. If the police did

1. To use the expression chosen by the philosopher Frédéric Gros for the title of his book.

not deal seriously with the troubles in the suburbs, these troubles would waste no time in spreading to our more affluent areas. Thus re-establishing peace and ensuring that it is maintained seems to be the long-term strategic prospect for Western armed forces. Since it is easier –and in the long run more profitable– to reinforce states than to support peoples, given that we know what the collapse of a state can mean in terms of massacres, the wars we are likely to fight will involve changing a regime or reconstructing a state. In both of these cases, what is important is not the planning of the war, but rather the question of the new order that is to be established and of the state that is to be instituted. It is therefore on these goals that the lines of action must converge, with military action being only one of many. This action will consist of a number of phases, including what the American armed forces call, incorrectly, Phase IV, "stability and support operations"; incorrectly, because it begins before the operations and is truly at the heart of the success of the intervention. Since the objective of such a war will be to establish, or re-establish, a reasonably stable government, the "victory" will not be a military victory. The weapons falling silent will be only a prelude, albeit an indispensable one. The military destruction of the opponent –which is today, wrongly, the prime object of the attention of military staffs and their attempts to achieve a transformation– will be only one milestone in the long line of operations. The final victory will be the one that counts: the stability of the new government. All the battles leading up to this must be seen in reverse order, without ever losing sight of the ultimate aim of the intervention: a self-supporting state of peace. In tomorrow's war, political planning is far more important than military planning.

In this reversal of interventions needed to reconstruct the state, we can see the relevance of the "trinity" of war. Its success depends on solving the written equations –both for the "intervener" and the "intervenee"– using Clausewitz's three variables of the state, the population and the army, and their three associated dimensions of rationality, irrationality and contingency.

A growing internal dimension

Today, globalisation has brought about a change in the age-old distinction between the internal and the external: "threats are no longer subject to borders". This is particularly true in Europe, where the opening of borders is inherent in the European idea. Consequently, the boundary between internal and external security has been blurred and the model of inside and outside forces –conceived by the Comte de Guibert on the eve of the Revolution and going on to affect democratic thinking through the two centuries of its application– has become obsolete. In this way, such apparently diverse deployment possibilities as a Western conflict, a stabilisation operation on the fringes of Europe or an outbreak of terrorist violence on national territory call for a comprehensive examination of the ways in which the same military capacity can be used in such widely differing circumstances.

Any description of the new context of engagement would therefore be incomplete if it were not to include the internal demands which lead armed forces –in particular, armies– to take part in collective security actions. The first duty of a state is to defend and protect its territory and its population. It must therefore retain the means to carry out effective action at home, which implies, amongst other things, retaining sufficient manpower to be able to carry out large-scale humanitarian action, for example, to prevent terrorist action. Increasingly, armed forces must be capable of operating at home, providing help to the population and, if required, responding to acts of terrorism. The essential operational function, which corresponds to the desired end effect, is the protection of national territory, including overseas territories, and our nationals.[1] Any other functions must be conceived in a way that allows them to contribute to this objective.

For the present, it is primarily within the framework of national solidarity, and of providing help and support to populations, that countries benefit from the manpower and unique capabilities of the armed forces, in situations where the police or other state services are no longer able or do not have the capacity to intervene. In these public service missions, in response to natural, technological or health disasters or in support of individual, but recurring, major world events, where their excellence can complement other services, the armed forces make a valuable contribution through their technical competence, rapid reaction capability, organisation and ability to remain solid in a crisis, born of experience. The occasions on which the armed forces are likely to be called upon to act on behalf of the population are unfortunately liable to increase given the population's great sensitivity to such disasters, which are also increasingly likely to occur. It is also possible that growing criminality and the creeping proliferation of areas in which law and order have broken down may lead to new forms of intervention on national territory, taking account, however, of the specific military character of the force, that is in itself a guarantee of the durability of the nation.

---

1. It is well known that the protection of our nationals goes far beyond the borders of France. It is perhaps interesting to note that the sixth largest French city is London, where over 300,000 French citizens have made their home.

In addition to these special forms of engagement, the armed forces, of course, already play a role in protecting the population against terrorism. Having been able to maintain their numbers and their functional autonomy, they remain the last resort of the government, able to react rapidly when everything else has stopped functioning. Thus they are always able to deploy very rapidly, and in large numbers, on national territory, to prevent an act of terrorism or secure a area already hit. As the old nuclear order no longer confers the privilege of immunity in the face of new forms of threat, the day-to-day risk is, sadly, no longer the prerogative of others; it is thus very much to be feared that these occurrences will increase, entailing the need for corresponding numbers and specific training. The porous nature of national borders is a major advantage to the enemy who has no territory, making it essential to have, on national territory and in the context of an increasingly close relationship between civilian defence/security and military defence, soldiers trained to tackle "grey" defence and crises, both malevolent and accidental.

The population would not understand, and with good reason, if the military instrument were unable to play a role in a domestic context. This situation would be dangerous. The English expression is quite clear: "If you are not useful, then you are not affordable"; there is no point in hanging on to a tool that is of no use. In other words, expensive and high-tech defence assets that are unable to deploy in the numbers needed to deal with, for example, a natural disaster, are condemned to gradually give up significant portions of their budget.

If the armed forces were no longer to be seen as a tool of direct benefit to our fellow-citizens, they would be in danger of no longer being able to conduct the essential military actions required in a front-line battle. Their budgets would shift inexorably towards other ministries, which would, at the same time, deprive the government of the ability to conduct actions outside the country. In the words of the Chief of the French Army Staff, General Bruno Cuche, modern armed forces must be "dual" forces, able to conduct both operations outside national territory and direct action for the benefit and security of the population. This is a powerful idea, that must be implemented rapidly, in a way that has all-round support. We need to establish the paths that will enable our forces to operate at home and abroad, with the same men and the same assets. This may well be the greatest challenge the armed forces will have to face in the immediate future, in a configuration rendered very delicate by the importance of political issues, as demonstrated by the attack in Madrid (11 March 2004) and its operational and international repercussions.

CHAPTER III

THE NEW ADVERSARY

Since the general framework and the political objectives of the use of force have undergone a profound change, the modes of confrontation have also altered their "grammar". Emerging from the breakdown of states, so not bound by state constraints, new forms of conflict are now giving free rein to internal dynamics which are difficult to control.

With its increased autonomy, the new violence is once again entering the private domain. It directly benefits from the fact that weapons of destruction have become more commonplace and more widely distributed. They are no longer solely the province of the strong and the rich. Without exception, war is losing its traditional role as a of means of communication between two states.

We need to accept that we now face a new opponent: refusing to understand this today will mean learning it tomorrow, to our cost.

A REGULAR INFRA-STATE ADVERSARY

The disappearance of the inter-state war leaves the field open to extra-state actors and –unbound by the clearly defined codes of a confrontation between two blocs– gives a new importance to the emotional and the irrational in strategic conduct. These are transient, ephemeral actors, difficult to identify and unorganised, whom traditional armies are tackling with modern weapons, often ill-suited to the ancient techniques of massacre. The granularity of the organisation of violence is being continuously refined: the enemy consisting of major units has been superseded by an enemy broken up into small cells, whose lack of capability characterises the decline in strength. By adopting this dispersed configuration and no longer exposing his centres of gravity –the traditional target– to precision strikes, the enemy is able to use force with an increased level of impunity. Unable to pursue directly major strategic and political aims, he achieves them through a succession of low-level tactical actions, thereby shielding himself from weapons supposedly capable of achieving rapid strategic success, but ineffective in these new contexts. The enemy's aim is to attack the vulnerable areas of Western societies, that is to say, their human dimension, while himself having access to a resource limited even further by embracing the cult of suicide as a means of strengthening his advantage,. France is therefore faced with a strategy –from "weak to strong"– whose virtues it is familiar with and which, by linking its own destruction to that of the opponent, shows its contempt for the superior power.

"Irregular" forces are only irregular in relation to their status as a belligerent party and to our Western ideas. If irregular conflicts using all forms of violence in an unrestricted way become the norm, the "rule" will no longer be relevant. The opponent we face in tomorrow's wars will in fact be "regular", because we will be encountering him "regularly". This makes it possible to satisfy those who support traditional war; they can rest assured that tomorrow's war will be neither regular nor irregular, as with each passing day the boundary between the two becomes more blurred. Free and creative, caring little for our standards and our codes of conduct, the opponent will be regular in his irregularity with regard to "our" rules, our customary conduct, our moral code and our Western codification of war. Irregular warfare, the sort of war we are likely to face –which has, in fact, always co-existed alongside regular warfare– is still war; it is certainly not an "operation other than war". This misunderstanding and this dichotomy have led, and continue to lead, to grave errors.

War is war; an irregular war is at present –and for the foreseeable future– its most likely manifestation. The grammar adopted in the past has not really been affected by technological developments. As observed by US Defense Secretary Gates[1]: "History shows us that smaller, irregular forces –insurgents, guerrillas, terrorists– have for centuries found ways to harass and frustrate larger, regular armies and sow chaos. [...] The toys and trappings have changed, but the fundamentals have not." The opponent will apply the two basic rules for survival in war: continuous adaptation and surprise. Just as we will endeavour to do, he will force us to fight on his terms –we have the example of Hezbollah in the summer of 2006, which forced the Israeli Army to fight in the style and place of its choosing– or, conversely, his provocative tactics will oblige us to deploy our weapons, designed for industrial warfare, in heavily populated areas, with disastrous consequences. The only way to avoid being surprised by the so-called unexpected is to know that it is certain to happen and that, far from being an

---

1. Speech on 10 October 2007.

impersonal planning goal, the opponent is a living being, who is constantly reacting to us and attempting to foil our plans, rather than trying to conform to them, and establish a strategy whose primary purpose is to foil ours.

Shaped by a culture favouring bold, rapid incursions deep into enemy territory without worrying about the flanks, in July 2006 the Israeli Defence Forces (IDF) were surprised by a guerrilla force prepared and trained for this type of manoeuvre, foiling it by means of a succession of braking and dodging ploys, together with a multitude of "swarming" actions, in other words, by a neo-non-battle.[1] Having stolen the initiative from the army whose very trademark it has become, Hezbollah managed to retain it, at both strategic and tactical levels, and to force the State of Israel and its forces into a permanent reactive position. Who can forget the pained question asked by an Israeli reservist in August 2006, just a few days after the end of open hostilities: "Why did we fight the war that Hezbollah had prepared itself for?"[2] It is of course very tempting to try and impose one's own law on the enemy, to oblige him to fight "our war". However, it is more realistic to remember that, as the Americans say, "the enemy gets a vote", something which has come as a painful realisation to Western forces in recent interventions. This is precisely what US Defense Secretary Gates went on to say in the same speech: "Our enemies and potential adversaries –including nation states– have gone to school on us. They saw what America's technology and firepower did to Saddam's army in 1991 and again in 2003, and they've seen what IEDs are doing to the American military today. It is hard to conceive of any country challenging the United States directly on the ground – at least for some years to come."

The good news is that soon there will no longer be any asymmetric enemies. The opponent will continue to avoid any head-on clash with a high-tech force, minimise the comparative advantages of such a force and exploit its weaknesses, particularly its psychological weaknesses. However, there will no longer be an asymmetric enemy, because we will soon come to realise that the reason we describe an enemy as "asymmetric" is that, intellectually, we are still stuck in an industrial way of thinking. We are obsessed by this very French idea of "strong to strong" and have not yet grasped the notion that any victory is asymmetrical. The use of the term asymmetric –whether it is used to describe technology or a moral code– reflects the refusal to accept that an adversary worthy of the name would wish to fight according to any principles other than ours. When today we speak of the "rules of war" we still mean "the rules of our war". Before long, however, we will come to

---

1. Guy Brossolet, *Essai sur la non-bataille* [Essay on the non-battle]
2. Joël David, *La colère des réservistes de Tsahal* [The Anger of the IDF Reservists], La Croix, 24 August 2006.

understand that our opponent is not an asymmetric opponent, but simply an opponent, even if he is "irregular". We will have understood that war has not changed; it is still essentially a duel. Furthermore, we will also have understood that, as a point of principle, an opponent will not comply with the rules and that there will be little point bemoaning the fact that he has obliged us to fight in civilian areas, that he does not wear a uniform, that he is using what we would qualify as terrorist tactics... We will also have understood that the very idea of an asymmetric war is anomalous, as it forms an obstacle to constructing a military instrument suited to the nature of today's conflicts. In the meantime, we will need to be cautious, ensuring that we do not envisage the conflicts of the future, using the equipment of the future, but against the enemies of the past. It must be clear to us all that a political victory may be achieved through the absence of a military victory.

AN ADVERSARY WHO IS NOT LIKE US

Our Western rationality leads us to "organise" the opponent in the same way as we do ourselves, even if we are no longer talking about a state as opponent. In other words, we see him as a system and attack him as such, according to the principles devised so long ago by the likes of Giulio Douhet, William Mitchell, John Warden and David Deptula. However, the opponent's intelligence –or possibly his nature– means that he organises himself differently, adopting intricate structures able easily to resist our strikes on his non-existent vital centres. This is the problem behind the fight against Al Qaeda, and those committing terrorist acts in its name, or of the Israeli Army's confrontation with Hezbollah during the summer of 2006. Hezbollah, far from being organised as a system, had opted for the American concept of a "distributed network", consisting of small, highly independent teams able to adapt rapidly to local conditions. In doing so, it made it impossible to apply the principle of the OODA (Observation, Orientation, Decision, Action) cycle now crucial to the structure of Western forces. The Israeli Army was unable to interfere rapidly in Hezbollah's decisional loop quite simply because no such thing existed. In the absence of clear "enemy systems", "nodes" or "centres of gravity", a high-tech manoeuvre is unable to produce a shock to the system. Rupert Smith[1] refers to a "rhizome" enemy, whose command system is very difficult to attack, just as rhizomatic weeds are difficult to eradicate. He believes that there are three possible methods, none of which corresponds to industrial warfare: discourage them, poison them with a systematic weed killer or remove the nourishment from the soil. In this sense, air power has become the

---

1. The Utility of Force.

victim of its own success. As Etienne de Durand puts it: "By making it very difficult for exposed enemy concentrations to survive, air power has largely succeeded in 'emptying' the modern battlefield, thereby displacing the opponents from developed countries to asymmetric forms of combat, in various degrees. In other words, from the rejection of the battle to the rejection of *jus in bello*."[1] By the same token, the Western concept of network-centric warfare becomes significantly less important.

The other interesting characteristic of the opponent is that he is often multi-faceted. Other than during the brief initial phase of the intervention, during which the enemy may attack as a single entity, an opponent can very rapidly comprise numerous independent elements pursuing different goals and only rarely organised into some form of system. We therefore frequently have to respond to a collection of only vaguely correlated non-systems, which excludes immediately any simple or centralised solution.

AN ADVERSARY WHO IS NO LONGER THE IRREGULAR ADVERSARY OF THE PAST

When faced with the problems caused by our new opponent, we frequently turn to the military writings shaped by decades of fighting the Maoist-style revolutionary enemy of the subversive revolutions of the Cold War. This is good, provided the differences are clearly understood and we do not simply apply the procedures adopted in the past. "Traditional" insurrections, whose purpose is to bring down a government by military means and establish their own regime in a vanquished capital, have become increasingly rare: at best, taking control of a limited area –as in Columbia, Sri Lanka, Iraq, India, etc.– satisfies ambitions which do not include taking overall responsibility. The central command of yesterday's revolutionary opponent, with its systemic structure, has been replaced by a faith, an identity and a feeling of belonging, which are enough to enable actions to converge on a distant objective, while avoiding forming any kind of strategic target. The use of information technologies and network-based working do away with the need for a vulnerable systemic hierarchy. We are thus deprived of the "centres of gravity" so beloved of our way of operational thinking. How ridiculous the pack of 55 playing cards bearing the faces of prominent Iraqis, that every GI had in his pocket in 2003, now appears! Then, the idea was that eliminating all these figures would be enough to complete the mission, especially once the weapons of mass destruction had also been found – concrete proof of the "just war".

---

1. Etienne de Durand, *Les faces cachées de la puissance aérienne* [The hidden face of air power], *Revue de la défense nationale* [National Defence Review], June 2007, p. 29.

In addition, today's opponent depends less on the local population for information, funding and support, and his need for troop numbers has declined. He therefore presents few targets that can be attacked by our modern weapons, which in turn are no longer able to demonstrate their effectiveness by disrupting supply sources and routes, as these have become less crucial for our opponent. Such absence of a "strategic" enemy means that we must destroy cell by cell, and reinforces the importance of tactical acts for the overall strategic manoeuvre. The large-scale manoeuvres and firepower of the industrial war have only a limited effect against small hostile groups acting independently.

Even more than the revolutionary enemy of the guerrilla wars of decolonisation, our probable future asymmetric opponent can only be beaten by asymmetric methods. More than in the past he rejects the fight or, if he somehow accepts in error, he rejects the outcome. The only "military" method of defeating this kind of enemy is to clear an area and hold it indefinitely, closing off all access. This is the tactic of "clear, hold and build"; although ideal in theory, in practice it is costly in terms of personnel, for an uncertain result. History demonstrates that this tactic has only ever been successful for a short period, and using methods considered unthinkable today. The only means of achieving political effectiveness in a different context has been "pacification" based on a completely different model; re-establishing "normal life" by the sword. Thus, it would appear that it is not possible to fight asymmetry directly, but rather that it is preferable to adopt an indirect approach, acting in parallel across all lines of operation, in particular the non-military lines. Wars against political and ideological enemies cannot be won simply by attacking their armed forces; direct steps are doomed to failure.

AN "UNRESTRICTED" ENEMY

One of the greatest problems relating to tomorrow's wars is that it is only from our point of view that they will be "restricted conflicts". There is no reason why we should agree some common rules of moderation with our opponent, since in all likelihood we will have a widely differing perception of the Other Side. We will have no hostile intent towards the Other Side as such, but rather a political intent towards an "object of war". The Other Side, in contrast, may see us as his "sworn enemy" and begrudge us our survival if it comes at the cost of his own. Thus, considering the two antagonistic positions to be irreconcilable, he may feel that all methods and all opportunities are legitimate.

For us, the value of the political aims at stake will undoubtedly be restricted –leading us to adopt a "restricted" approach– whereas for the

Other Side, they will often have an absolute value. The scope of the issues motivating him –an exploitative religion or simply survival in the war of the dispossessed– will cause him to adopt what we term the logic of total war. In other words, war characterised by radicalism, the absence of limits between what can be considered as war and what not, the absence of limits in relation to aims, since they are expressed in absolute terms, together with the absence of limits in relation to resources, and even, with regard to time. The resentment felt by the fanatic, the disregarded, the weak or the misunderstood leads to the radicalisation of irregular behaviour. It leads to hate or, in other words, to excess. Clausewitz observed that it is symbolic stakes that lead to total war and, from one side at least, there will be no shortage of these in tomorrow's war.

This dissymmetry in approach makes things tricky for Western powers intervening in conflicts limited only from their point of view, while the opposing party sets no constraints on its action. This was the case in most wars of decolonisation, such as the Vietnam War in which, "the North Vietnamese, under Ho Chi Minh and Giap, were prepared to exceed all limits, in terms of sacrifice, of place and of time."[1] It is also true of the majority of contemporary crises, such as Iraq, Afghanistan and Lebanon in the summer of 2006. Michel Goya[2] quite rightly speaks of the new paradigm of "total localised war", recalling the "Iraqi, Afghan, Palestinian or Lebanese Shiite rebels who, while conducting a total war, use 'total' methods, such as suicide combat." In parallel with this "totalisation" of combat, there is a visible and rapid decline in the technological and financial obstacles to access to extreme violence. The British call this phenomenon HICOIN (High Intensity Counter-Insurgency). This intensification of war –a natural and irrepressible movement linked to our superior strength, which must be circumvented, to the spread of technology and to the radicalisation of the aims of war– is growing stronger with each passing day. The "militant combatant"[3], a participant in the total localised war, sees his tactical and lethal effectiveness strengthened by the spread of "user-friendly" advanced technology, in conjunction with ideologies extolling individual sacrifice.

In the long term, firepower itself can have little effect against such an opponent. He is better able to resist losses than the intervening countries. For a wide range of reasons, linked to demography and aging populations, Western armies and, even more so, Western public opinion can accept only very modest losses. In contrast, opponents drawn from societies with

---

1. Carver, in Paret, p. 787.
2. Michel Goya, *Dix millions de dollars le milicien* [Ten Million Dollars per Militia Member], in *Politique Etrangère* [Foreign policy]., 1/2007.
3. Expression used by Jacques Sapir.

a strong demography and young populations are prepared to accept a bloodbath. The figures are striking, since in asymmetric wars the average ratio of losses between the two camps is 1 to 8^1. The future opponent will thus be able to withstand the physical attrition that we are likely to inflict on him for much longer than we will be able to put up with the ideological, media and political attrition to which he will subject us. The restrictions (geographical, political, ethical, etc.) which we will be obliged to impose on ourselves in future wars may even make our intervention pointless. The advance analysis of the real chances of success, despite our self-imposed limitations, will be of particular importance. Our future wars will always be difficult, since these constraints will make it hard to respect the principles of conventional war, which is our true area of expertise. Far removed from the passion of events, the rule seems quite clear: if the constraints we have to impose render the use of any military strategy ineffective, then the use of the military instrument to achieve political ends is unreasonable. US General Chilcoat expressed this truth very clearly: "If centers of gravity, the most vital military targets, lie beyond the political constraints imposed by the nation's leadership, military intervention is unlikely to succeed."[2]

AN ADVERSARY WHO IS ADAPTING INCREASINGLY RAPIDLY

According to the inherent rules of war, belligerent parties change their modes of action as circumstances dictate.[3] Our first problem in future wars

---

1. Algeria: 30,000/250,000 – Vietnam: 58,000/1,000,000 – Afghanistan (USSR): 13,500/50,000 – Iraq: 3,800/1,500,000 (end2007).
2. Richard A. Chilcoat, Strategic Art: the New Discipline for 21st century Leaders, Strategic Study Institute Publication, 10 October 1995.
3. On 3 September 2007, the Herald Tribune published a particularly interesting article on this subject: "A short-lived victory over the Taliban". It sets out the way in which the Taliban, having learnt the lessons from their defeats by the Coalition in 2006, completely changed their modes of action in 2007. In particular, they avoided setting themselves up as a target for the full strength of the opposing force. "A year after Canadian and US forces drove hundreds of Taliban fighters from the area, the Panjwai and Zhare districts southwest of Kandahar, the rebels are back and have adopted new tactics. Carrying out guerrilla attacks after NATO troops partly withdrew in July, they overran isolated police posts and are now operating in areas where they can mount attacks on Kandahar, the largest city in the south... The Panjwai and Zhare districts, in particular, highlight the changing nature of the fight in the south. The military operation there in September 2006 was the largest conventional battle in the country since 2002. But this year, the Taliban are avoiding set battles with NATO and instead are attacking the police and stepping up their use of roadside bombs, known as improvised explosive devices or IEDs... a 20 percent increase, according to the United Nations... After moving through the area in large groups last summer, the Taliban now operate in bands of no more than 20. Instead of sleeping in freshly dug bunkers and trenches, they sleep in mosques and houses, apparently to avoid NATO air strikes, or, in the event of an attack, to cause civilian casualties."

will be that the opponent will adapt even before the intervention commences, using his knowledge of the situation and of our forces to thicken the "fog of war" and reduce the effect of our best technological resources. The Serbians prepared well for the modes of action they expected to encounter: during 78 days of bombing, NATO destroyed over 500 dummy targets for only 50 real vehicles. Four years later, the Iraqis, ahead of the expected US initial decapitation strike, emptied their command posts and communications relay stations and installed their resources and networks in dozens of nondescript civilian buildings. The Shock and Awe attack struck the concrete, but made little impression on the minds of those who had prepared for it. It produced no confusion, no paralysis and no decapitation. For his part, the Hezbollah commander had understood well before 12 July 2006 that weapon systems which are not mounted on vehicles have a very low signature and are difficult to detect, track and destroy, yet easy to hide and move. The asymmetric stealth thus created foils the detection capabilities, which rely solely on technology. This tendency negates the effectiveness of the current concentration on "time-sensitive targets" as we attempt to respond to the specific difficulties of counterinsurgency operations. The American chronicler Ralph Peter described this problem clearly: "Not a single item in our trillion-dollar arsenal can compare with the genius of the suicide bomber – the breakthrough weapon of our time. Our intelligence systems cannot locate him, our arsenal cannot deter him, and, all too often, our soldiers cannot stop him before it is too late. [...] All of our technologies and comforting theories are confounded by the strength of the soul ablaze with faith."[1]

Our second problem is that, as well as adapting prior to the intervention, the opponent is increasingly able to adapt during the operation. Our hierarchical systems, based around a top-down approach, are naturally far less capable of a rapid reaction than those of the opponent which are organised in networks, with the individual cells having a large degree of autonomy and able to "fit" the circumstances far more easily. The Americans and the British are quite clear about their misadventures in Iraq: in spite of all their technology and knowledge they are always one stratagem or one trick behind the enemy. An official British document observed that during a single six-month tour by a brigade in Iraq, the improvised explosive devices planted roadside verges had been detonated in turn by radio remote control, interruption of an infrared beam, dual infrared control, breaking a wire and, finally, by a mobile on-board vehicle device! The opponent is taking advantage of virtually unlimited access to many of the most modern technologies

---

1. Weekly Standard, 6 February 2006.

and devising ever more sophisticated combat procedures. As the British themselves say: the Iraqi enemy has adapted far more in three years than the Irish enemy in thirty. The worst thing is that the opponent is evolving, but in a non-linear fashion, able to vary, in a way that is impossible to predict, the intensity and the mode of violence, leaving us bogged-down in hybrid wars where each one is different. At each stage of the evolution, we endeavour to provide a response, but the opponent continues to pose new questions.

Both reason and experience demonstrate the futility of striving to plan the opponent's reactions or spending time improving concepts based on his predictability. Like any other biological system, the enemy's system is too complex to be predictable in any meaningful way. Similarly, there is little point in searching for perfect planning or strategy, as to do so would be to fail to understand the fundamental law of war, that of reciprocal action. Thus, what counts is flexibility, adaptability and the ability to react to changes in circumstances. Furthermore, it is important to understand that solutions cannot be purely technical, quite simply because these are beyond the means of our economies. This is clearly illustrated by the budgetary effort –already referred to– required from the US Army simply to provide a technological response to the IED threat alone. Lieutenant colonel Bill Adamson, one of the project leaders, explained the problem: "They are adapting faster than we can acquire the technology."[1] In this area too we will need to start thinking differently.

In summary, with regard to our opponent there are four major risks. The first is simply to consider him as an insignificant factor in the analysis.[2] The second is to see him as another one of us, and so to ascribe to him our way of reasoning, our way of doing things, etc. The third is to treat him with contempt, which could easily happen since he does not have the same elements of strength as we do, which is precisely his strong point. We recall the American studies carried out in the euphoria of the fall of Baghdad in 2003, and which predicted that all wars would be like the "Indian Wars"![3]

---

1. De Defensa, 10 April 2006, p. 18.
2. This was probably one reason for the IDF's difficulties in July-August 2006. Pierre Razoux, a military historian specialising in the Middle East, who has written several books on this subject, wrote in a note from *IFRI* [French Institute for International Relations] entitled "After the fall – the re-orientation of the Israeli Defence Forces": "The Chief of Staff, Dan Halutz, probably thought that a new war in Southern Lebanon would be a unique opportunity to test his strategy, concentrating on remote operations, which would even make it possible to increase the role of the Air Force within the IDF." We know that the reality of this war was quite different and that the later re-orientation of the IDF favoured the land forces. (Study dated October 2007, collection *Focus Stratégique*)
3. Unfortunately, in a reversal of this, some of today's battles are more reminiscent of the famous and bloody defeat at the Battle of the Little Bighorn in the Black Hills of Montana/ Dakota (25.06.1876), where the Cheyenne and the Sioux, under the command of Chief Crazy Horse defeated the 7th (US) Cavalry led by Lieutenant Colonel George A. Custer.

Such contempt implies the rejection of the intelligence of the opponent and the refusal to accept that he is a better innovator than we are, or the rejection of the opponent's creative willpower. It also implies the overall simplification of entities that do not fit in to logic or diverse identities; confusion and the rejection of specific elements do not allow for the intelligence of crises. Contempt leads to caricature, that prevents us understanding properly, and thus fighting properly. Contempt is the best recipe for failure. The fourth risk is the "virtual reality" resulting from the digital requirements of modelling. Tomorrow's wars will be fought in the real world, not in an ideal world in which nations dictate rules and conduct, nor on video screens which give a false idea of what is feasible. Winning the war means seizing, and holding onto, the initiative; the problem is that, for the present at least, in the face of asymmetry we seem to have lost it.

CHAPTER IV

THE NEW LOCATION FOR WAR

The location of war has changed. In the past, wars were fought by armed forces in three dimensions and in open spaces. Those days have gone. Tomorrow's war is likely to be fought somewhere where the surroundings will have a levelling effect for Western superior power: in highly populated areas, on the ground, in confined spaces.

The "contested zones"[1] will be the contestable areas of the "grey zones"[2], those areas in which law and order have broken down, both internal legal order and international legal order, and where –as soon as control is relinquished– there will be a proliferation of forms of violence that will escalate and threaten us if we do not manage to contain and reduce them "in advance".

WAR AMIDST THE POPULATION[3]

Tomorrow's warfare will be fought not between societies but within societies. The population is thus automatically a major actor and a major issue. There are no conflicts in which the civilian population is not central to the military concerns of the parties involved. Moving from a world in which the population represented the "rear area" –as opposed to the front, the essential military zone– armed forces now operate in its midst and with reference to it. The armed forces have entered the era of the war amidst the population.

---

1. Expression used by B. Posen, "Command of the Commons", International Security, vol. 28, no. 1, June 2003.
2. For those responsible for the control and safety of aircraft, the "grey area" is a set concept. It applies to the whole area of airspace with no, or only poor, radar coverage. Security specialists have borrowed the term from this technical discipline.
3. General Sir Rupert Smith's book, The Utility of Force, provides very interesting reading on this subject.

In a confrontation with Western armed forces, these populations can be easily mobilised to form a hostile crowd: an "army" that we cannot destroy and a force which can operate behind a protective mask. However, just like any other opponent, they are the target that we have to influence and an objective whose reactions will determine whether the final result is success or failure. Since this objective consists of human society, with its governance, its social contract and its institutions, rather than a province, a river or a border, there is no line or no area to be conquered or protected. The only front to be held by the forces involved is that of the people. Thus, in order to be effective, the use of forces must be aimed at achieving a political effect akin to that which the populations, plunged into disorder, chaos and uncertainty, are hoping for. This essential requirement, which if disregarded would mean failure, imposes considerable constraints on the use of forces and on the definition of objectives. It also introduces a significant ethical dimension into conflicts where a psychological confrontation represents an important part of the result.

A war fought amidst the population seems to be far more complex in nature than an industrial war. The technical problems remain, and are, in fact, even more complicated, but in addition there is the considerable complexity of the human environment. An insurrection cannot be dealt with using the same weapons and methods used in the past to destroy a column of Soviet tanks. War and peace within societies is not the same as war and peace between states. Interference, the key characteristic of intervention, involves an external actor changing an "internal" situation by means of force; it upsets the balance and provokes complex and unpredictable reactions. At its heart lies an inequality between "intervenee" and "intervener"; however laudable the motives for the intervention, it is initially a source of resentment and frustration.

Commitments toward the population, the restoration of economic and political life and the legitimacy of the action are all reasons for ensuring the proportional use of force. Destruction cannot be excluded: it is an important option available to us, especially since the threat of using lethal force often acts as a deterrent or influences the course of the action. Force must therefore be credible –this is the first lesson that can be learnt from the difficulties encountered by UNPROFOR in Bosnia in 1995– that is to say, it must be of a form and scale to act effectively and to respond forcefully to attempts to disrupt a return to security. This is probably one of the strongest recommendations contained in the Brahimi Report[1], which discusses the lessons learned from the difficulties encountered by the UN

---

1. UN, 21.08.2000.

forces: "Peacekeepers must be capable of defending themselves, other mission components and the mission's mandate against those who seek to undermine it by violence. [...] No amount of good intentions can substitute for the fundamental ability to project credible force." Of course, the use of force must be measured, accurate and confined to the objective. It should limit, and if possible exclude, collateral damage, while remaining firm and decisive. This question of credibility is crucial, since deterrence is first and foremost a matter of perception. The credibility of the intervention is based primarily on force, firepower and the dynamics of victory. The sense of an irreversible advance is a key element in winning over the population, since as far as the former masters are concerned, siding with the opponent means signing one's own death sentence.[1]

For the military in the early 1990s, the lessons learned by the "blue berets", full of false hopes born of the euphoria that followed the fall of the Wall, were painful ones. Unfortunately, we had a tendency to confuse the end (peace) with the means (force), and force is necessary to rebuild peace. The expression "peacekeeping forces" reflected a pernicious and dangerous view of forces and of peace. Forces are not "peace forces"; they are forces of war, but are nevertheless an important tool in achieving peace, and should be honoured to be part of this. We will not decide which crises to get involved in, nor which battles we will be called upon to fight; however, in both cases, we must always be capable of "winning", even though this word does not fully do justice to the complexity of tomorrow's war.

These various, seemingly contradictory, requirements affect the choice of weapons, whose effects must be graduated in order to take account of a wide range of threats and actions. We must be able to deter, to respond to an equal opponent, to control an unarmed crowd and to have an alternative when we find ourselves facing child-soldiers. It must also be possible to change our position immediately, whether it be to reinforce our attitude when violence threatens to get out of control or, conversely, to return to a more peaceful situation.

THE LAND ENVIRONMENT – MORE ESSENTIAL THAN EVER

Since force is primarily used between human societies, it occurs principally on land. It is on the ground that crises build up and are resolved.

---

1. In *Atlante-Aréthuse: une opération de pacification en Indochine* [Atlante-Aréthuse: a Pacification Operation in Indochina], Michel Grintchenko offers an interesting study of this subject.

While future wars may start remotely, from the sea or from the air, they will always end on the ground, and will eventually involve the local population and a direct confrontation with the other side. It is on the ground, where there is physical contact between the protagonists, and a lasting presence, that the action will bear fruit.

When it does not entail the total destruction of the opponent, war has always been about controlling the opponent. This is because war primarily means imposing one's will and such imposition is not possible without control, no matter what means are involved. Today, even more than in the past, because the target of the action is the population rather than the state, winning a war means being in control of the environment. In the case of a war being conducted in the midst of the population, against what can only be an irregular opponent, the only method of exercising control is on the ground, where the population and the irregular opponent live. Thus controlling the environment on the ground is the core and the essence of the "stabilisation phase" of a conflict, the "decision phase"; this is the phase in which the desired strategic effect is prepared. Such action on the ground makes it possible to move from a military objective to a political goal. It requires forces that can operate in the conflict zone for a prolonged period, and which are capable of recovering and adapting to changes in circumstances. The analyst Frederick W. Kagan is clear in his views: "The decision to develop a method of war that relies for success primarily on identifying and destroying targets, rather than occupying territory therefore reveals a fundamental misunderstanding of the nature of war itself. [...] The size of the ground force needed to control conquered territory is determined by the size of that territory, the density of its population, and the nature and size of the resistance, not by the nature of the soldiers' weapons. [...] As long as war remains a process of human beings interacting with one another –as all irregular warfare is– the land-power "market" will require a heavy investment in people."[1]

Given the fluid nature of new conflict situations, a good knowledge of these situations will be a key factor in the effectiveness of the force, as it is a crucial factor in enabling continuous adaptation to events. Since identification of the opponent will be uncertain, and distinguishing between friend and foe only possible with the help of human intelligence and contact, understanding the threat requires an "in the field" vision rather than a technical vision. It is no longer a case of detecting massed tanks and locating potential targets, but of understanding social environments, behaviours and psychologies. It is also a matter of influencing human wills

---

1. The US Military's Manpower Crisis, in Defense and Arms, 22 June 2006.

through the selective and proportionate use of strength. In the sort of war we are likely to have to fight, the commander will have to use tact and diplomacy to gain an insight into the delicate nature of a complex and changing environment. This is only possible on the ground, in direct contact with the situation. Armed with this knowledge, he will be able to regulate the level of forces, and to ensure that any steps taken can be reversed if the fluidity of the situation requires it. In doing so, he is thus able to preserve a continuous link between tactical and political action.

In the past too, only strength on the ground could be a decisive factor, even though the Duke of Wellington's victory was only possible thanks to Lord Nelson's earlier success; the victories in the Battle of Britain and the Battle of the Atlantic were necessary to ensure the fall of Berlin; and ships and aircraft were needed to enable and support the Marine Corps assaults in the Pacific Islands. But only the side that plants its flag on the enemy's territory is the victor; the ultimate test of wills is the direct confrontation between soldiers in face-to-face combat. Although the notion of victory has changed, the importance of strength on the ground will be even greater in tomorrow's war. The political success sought relies on effective support by the navy and the air force, but the decisive effect –the one which permits political success and thus the success of the intervention– will be achieved on the ground by the convergence of a variety of effects, only some of which will be military. At a political level, whether we are talking about relations with other partner nations or the population surrounding our intervention, it is the engagement of forces that represents the strongest and most important political commitment. It denotes the concrete aspect of political repercussions, particularly because this engagement is a weighty decision, with far-reaching implications. The deployment of land forces, and thus the price of blood, are the only real proof of political determination.

This major role played by the population, at the heart of interventions, means that the composition of forces can no longer be the same as when wars took place between states. It is now based far more on the ability to meet the opponent on his own ground, sticking as closely as possible to his changing reality. The traditional top-down approach of inter-state wars (attacking the state, focussing primarily on action in the third dimension) has been supplanted by the bottom-up approach, since it is so often a case of starting from the lowest level and working to rebuild the state. As a result of the distance from the terrain at stake, the Revolution in Military Affairs appears to be out of step with recent developments in conflicts and unable, in most cases, to provide an appropriate response. We will look at this later in more detail.

The Chief of the French Army Staff, General Bruno Cuche, passed on this message to the army cadets at the Ecole Supérieure de Guerre in June 2007: "It is the sum of the tactical actions, and no longer the centralised stand-off action from the sea or the air, that enables us to achieve the desired final result. Individually and collectively our soldiers today are the guardians of part of the strategic effect. The ground environment is now, and will be increasingly so in the future, the centre of gravity of joint campaigns because it is here that we find the population. Thus the intervention of the land forces is proof most positive of the political commitment."

Our probable opponent is watching us. He knows that it is only by engaging our land forces that we will be able to constrain him and prevent him from fulfilling his aims[1]. The superiority of these forces, in quantity and quality, is in itself a deterrent; it must be maintained to avoid the development of asymmetric conflicts over which we, as hostages of the ineffective stand-off strike, would have no control. The generic characteristics of the probable future war have one clear consequence: under present conditions, the capacities of the land forces play an essential role in the internal and external security of France. They form a major lever with regard to maintaining or improving France's position in the world, its ability to act and its ability to impose its political will. In tomorrow's war, the land forces will be the strength of the nation. With forces operating henceforth principally in a populated environment, the city will replace the countryside as the main area of action. Success in urban combat situations now defines the new utility of weapons. However, it is a case of winning not the battle for the city, but rather the battle in the midst of the

---

1. It is interesting to note that one of the most important decisions taken by Israel in response to the problems encountered by its technological, air-centric war in July 2006, was precisely to reinforce its land forces, following years of reducing budgets in favour of the air forces. "Following a year-long analysis, the general-staff established a five-year plan, the Tefen Plan, giving priority to the land weapons programmes in the budgets dominated for so long by aviation and the navy. [...] It reversed a decade of budgetary attrition and of downsizing; it corrected the budgetary imbalance which for years had strongly favoured technology over men. [...] The plan represents a return to the basics of combat. The most important aspect is an increase in land forces, with the creation of two new divisions. The intention is now to put the emphasis as much on numbers as on technology, which Israel now believes must not be funded to the detriment of operational preparation and training. Instead of the disappearance of tanks, included in the original plan for 2007-2011, the Tefen Plan envisages the deployment of dozens of new upgraded Mk4 Merkava tanks and hundreds of new heavy personnel carriers on the same chassis. According to the Vice-Chief of the Joint Staff, General Kaplinsky, the first priority of the Tefen Plan is a significant strengthening of the land forces, enabling them to achieve a rapid decision on the battlefield." Extract from an article by B. Opall-Rome, published in Defense News on 10 September 2007.

population of the city. As Thucydides wisely reminds us: "Men make a city, not walls."[1]

"You are a lucky man!", General de Gaulle is reported to have said to General Leclerc, on the eve of the liberation of Paris. But times have changed. War, the city and the soldier have been battling it out for millennia, but times have now changed. These days, commanders in charge of fighting in urban areas are much more likely to be fully aware of their responsibilities and of the complexity of their task, than to be feeling lucky.

Because they can deploy and operate more freely there, armies have always preferred open spaces to the confined spaces in towns, and their equipment, structure and training have been tailored to this. To a large extent, they still are. However, the soldier in the 21st century has little choice; currently, peace must be imposed and maintained in towns. Nowadays, without exception, combat takes place in an urban environment and the names of battles are the names of cities: Sarajevo, Grozny, Beirut, Baghdad, Mitrovica, Basra, Abidjan, Bint Jbeil... The route to peace now passes through the city. Ignoring this phenomenon does not change the fact of its existence. In the Iraqi Freedom campaign in March and April 2003, the initial choice was to go round the cities, but this was swiftly followed by an abrupt return to reality: modern wars are fought in cities, they cause pain, they involve close contact, they last a long time...

The increasingly urban nature (a splendid irony!) of operations is due to three essential reasons. The first is the evolution of the nature of conflict. 80% of the wars fought since 1945 have been internal conflicts or civil wars. War, departing further and further from the Westphalian order and becoming less and less an affair between states and more a civilian matter, flourishes in cities, at the heart of human society. The second is the vigorous phenomenon of urbanisation. Forecasts vary in their detail, but the general picture is the same: in 1950, 75% of the world population lived in rural areas. Today, there is a growing trend towards very large cities (with over 300 cities where the population exceeds 1 million) and half of the population of the planet is concentrated in cities and. By 2035, this figure will increase to 65%. These extremely large cities are devouring the spaces that surround them at an increasingly fast rate, often creating areas of uncontrolled disorder. The third reason is the concentration of power and wealth, the key issues in most human confrontations. Both are primarily to be found in cities, making crises most likely there. We can no longer devise theories and plans allowing us to avoid the city.

---

1. Thucydides, p. 530.

Five key ideas help us to reflect on urban warfare, and thus on the war we are likely to encounter.

The first of these ideas is the very specific nature of the urban environment. In a town the atmosphere is not purely military and rational. It is also –and indeed primarily– civilian and emotional, demanding a comprehensive approach. For the soldier, this translates into a heightened relationship with his environment, taking into consideration a wide range of aspects (cultural, political, legal, religious, humanitarian, media and, of course, military). The complex variety of actions involved requires close coordination with a broad spectrum of actors, including two new ones: non-governmental organisations and the media. Humanitarian workers and the military are obliged to operate in the same space; their actions must at the very least be coordinated, but cooperation is preferable. The media is always present in great numbers. Urban warfare is spectacular; while there may be no unity of action, it has the unity of place and the unity of time found in classical tragedy. Urban areas bring together the symbols in a confined space, concentrate the violence and focus attention; under fire from the media, they focus all the concerns and all the interest, thereby accentuating even further the increasingly litigious nature of military action and thus the restrictions on the freedom of military action when, in this complex environment, success comes through initiatives in the field.[1]

The second idea is the evolution of the opponent. Urban areas favour asymmetric combat since, by minimising inferiority, and thus relative differences in strength, they offer a safe place where the weak can confront the strong. Here the opponent can easily reject the Western view of war and, using non-conventional ideas, can attempt to attack the moral forces of the combatants and their nations, rather than their physical forces. For the opponent, this is a favourable site, where he can find much support. Assassinations, ambushes, harassment, sniper fire, IEDs, infiltration, disinformation and crowd movements: these are the weapons he deploys and which we must fight and protect ourselves against. It is in a city that an asymmetric soldier is best able, under the constant scrutiny of the media, to pursue his strategy of provocation and propaganda, aided by the collateral "blunders" of his opponent.

The third idea is the changes in the concept of military effectiveness. Nowadays, the problems to be solved are primarily human, rather than

---

1. We must fight against inhibition, the source of operational ineffectiveness and the chief consequence of the combination of the increased media coverage of conflicts and the extension of legalism to the operational sphere. In his speech to the Ecole Navale on 25 January 2006, the French Chief of the Joint Staff spoke of the risk of seeing a battlefield full of "inhibited lawyers in uniforms, with a law book in one hand and an umbrella in the other."

physical. The target of the action is no longer the opponent, but the population. In the heart of the city, it is a matter of winning the battle for popularity, while re-establishing the "social contract". This is achieved through a diverse range of urban missions (coercion, security, humanitarian aid) to be carried out simultaneously. Urban areas demand an all-embracing approach and the synchronising of actions.

The fourth idea is a direct consequence of the third: the evolution of the resources needed for military effectiveness. It is very costly to capture, hold and control a city. A city imprisons those fighting there, setting traps for those who long for wide open spaces and forcing them to act differently. The city, heterogeneous and opaque, sets the conditions. Forces organised into different components designed to win a rapid and brutal head-on battle in an open area prove unsuitable; the physical and human characteristics of the urban environment make it necessary to re-examine effectiveness. Tactical aspects are more important than operational aspects; decentralisation is more important than centralisation; influence is more important than strength, and individual action counts almost as much as a headquarters' decision and speaks much louder than the strategic message.

The use of force must be controlled and modulated, in order to maintain the conditions of a future "return to normality". Destruction must be kept to a minimum – a revolutionary change in attitude since, until the end of the Cold War, the ability to destroy was the measure of military effectiveness. Things have changed since 16 April 1945, when 40,000 howitzers were fired at Berlin simultaneously. In an urban environment, "non-kinetic" capacity is just as important as "kinetic" capacity.

In the longer term, contact on the ground is proving to be essential and short-range or even close, all-arms, low-level combat is taking on a more significant role. Conversely, the relevance of stand-off combat is in sharp decline. The dream of "fire and forget" is fading in the face of the clear need to occupy the area that has been made safe, metre by metre. Further down the line, we see the return of true risk, that is to say, the increased cost of engagements in human terms, thus the rise in political risk. The style of command needs to evolve, with large-scale actions and centralised deployment giving way to decentralisation and small teams. Intelligence has also changed: in a city it clearly has specific characteristics, with human contact being more important than technical procedures, and each soldier becoming a gatherer of fragmented, diverse, uncertain, yet essential, information. Six months into the campaign in Baghdad, US General Dempsey, commander of the 1st (US) Armoured Division observed that: "In an urban environment, three quarters of all actions are intelligence actions."

In a city, the most likely location for tomorrow's war, understanding is more important than knowing.

As time passes, certain recent technologies, which seemed destined to play a major lasting role in our forces' future capabilities, are now being called into question. The physical characteristics of the urban environment restrict the effectiveness of some technologies, while other equipment acquires a renewed effectiveness. This necessitates a degree of readjustment and reorientation; this need to adapt systems and equipment originally designed for a style of combat that avoided cities will be keenly felt well outside the confines of the land forces.

The fifth idea –a revolution from many points of view– relates to the expansion of the military profession. Today, the city represents of the duality of the social and martial roles of the soldier as operations comprise many aspects other than warfighting. Whether fulfilling an immediate need or ensuring long-term peace, or achieving by the use of force or influence a political result in the complex fabric of the urban environment, the military are having to act as coordinators and managers, in liaison with various local actors, such as the police, humanitarian workers, the administration, politicians, etc. Speaking in October 2007, US Defense Secretary Gates noted: "Army soldiers can expect to be tasked with reviving public services, rebuilding infrastructure, and promoting good governance. All these so-called "non-traditional" capabilities have moved into the mainstream of military thinking, planning, and strategy – where they must stay."[1] These weighty changes make it necessary to rethink the identity of the soldier: urban operations, the core of tomorrow's war, are causing a distortion between his past identity and the military profession today; Marshals Gallieni and Lyautey would no doubt consider this simply a sensible return to expertise, the sort of expertise which only recently was so well understood.

In summary, the city has become the place which symbolises both the increasing complexity of the military profession and its duality: the city requires us to re-evaluate the effectiveness of the armed forces. The main task remains to take, hold and control the city and, of course, its population, even though it is costly to do so, because it is one of the last areas where an opponent can hope to defeat or resist a modern army. We have to live with this idea: the wars we are likely to have to fight –as well as the ones we are unlikely to encounter– will include an urban element, or will even be primarily urban and, even more than when operating in open areas, controlling this environment will require a combination of different actions (coercion, security, humanitarian aid), both civilian and military, at all levels.

---

1. In a speech on 10 October 2007.

CHAPTER V

NEW OPERATIONAL SPECIFICATIONS

TWO COMPLEMENTARY AND COMBINED MODES OF OPERATION

In a crisis, the armed forces employ two modes of operation, which are complementary in their effects and combined in their execution. This, proven, intertwining renders obsolete the artificial separation between "high intensity" and "low intensity" conflicts, since today we see only conflicts which alternate in time and space between periods of variable intensity. Such conflicts require us to have in place forces with multiple and transferable capabilities, accustomed to rapid changes to their modes of action and execution. While the distinction between "warfighting" and "operations other than war" still upheld by some may reflect specific strategic cultures, or may serve to protect certain force models, it in no way corresponds to the reality of contemporary crises and their continuum on the scale of violence.

Put briefly, we must initially be capable of forcing an opponent to renounce his actions, by threatening to destroy him, or by destroying him if necessary; as time progresses, we must be able to overcome violence, requiring us to have powerful and protected forces, capable of deterrence and prevention. In reality, there is no separation in time, but rather a constant overlap in which the dominant factor may change. It is interesting to note that, coercive actions are characterised by high level political intervention –Head of State or force commander– while the political management of de-escalation operations, which require a detailed understanding of the situation as well as the permanent presence of contact troops to carry them out, is often played out at the lowest level.

Such observations demonstrate clearly the impossibility of having forces dedicated solely to peacekeeping. The soldier on the ground must be capable of excelling simultaneously across the whole range of military actions. On the other hand, moving arbitrarily from one mode of action to another demands adaptable forces, able to change what they are doing instantaneously. It also demands a highly effective command/intelligence system, able –through the shortening of the decisional loop– to deal rapidly with any type of situation. To carry out all these different actions well requires heavy forces to deter and destroy, combined with appropriate technological superiority and armour to provide protection and limit losses to an acceptable level, thereby making the deterrence credible and avoiding losing the support of national public opinion. Whether the task is to achieve or maintain a ceasefire, armed force and its various tools are indispensable.

It is easy to identify the new demands arising from the requirement for the soldier to be capable of two modes of action, and of switching from one to the other at a moment's notice.

THE NEW CONTINUUM

Traditionally, military action was concentrated in an often violent, brief phase of action, preceded and followed by political action. This sequence has been utterly outmoded by the reality of modern crises. History may well look back on the banner proclaiming "Mission Accomplished", slung high on board the USS Abraham Lincoln on 1 May 2003, as marking the inability of the battle alone to achieve the desired strategic result in tomorrow's war.

The one, intense phase of the battle has now been replaced by a succession of phases, both discrete in their intention and more or less concurrent in their execution. Thus, because traditional war is dead, the simple concepts of war and peace are no longer adequate for describing current conflicts. Since neither the notions of coercion and taking control of violence, nor of high and low intensity can be used to define the phases of operations, still less the operations themselves, a new way must be found to describe operations.

By its very nature, modelling involves a process of simplification, leaving it open to criticism by those who forget the need for synthesis, yet it nevertheless enables us to gain a better understanding of the situation. It enables us to get a better grasp on reality, even if it is unable to reproduce all the hypotheses. Thus, one can keep to the vision of French military doctrine and define the course of current and foreseeable conflicts as a succession of three phases within a continuum: intervention, stabilisation and return to normality. These three phases have no precisely defined boundaries and indeed partly overlap, but they do have distinct characteristics.

The first phase –intervention– is indispensable by definition, since it introduces the intervention force. It may be a simple deployment, but it may also involve the use of a significant level of violence in order to defeat an armed opponent and impose temporary order. In the latter case, that of a true armed conflict, for some of the time the military aspects will outweigh the diplomatic aspects. In its violent form, this phase will require the deployment of powerful, joint forces, able to bring into play a formidable capacity to destroy –and, if necessary, to scatter– in order to convince the opponent to renounce his intentions. Once engaged, the force must display its determination to take control, using appropriate means, or risk jeopardising its credibility. Failure to "win" would mean endangering the whole intervention process. In most cases, however, the aim will not be to abolish the existing political order. Backed up by a good knowledge of the area, it will be more a case of changing it in order to make use of it, and thus to avoid it disintegrating into a myriad sources of rebellion far more difficult to deal with in the subsequent phases. This intervention phase is, in principle, a phase characterised by a clearly defined objective: a military victory on the battlefield, even though, in the new context, the aim is to neutralise rather than to destroy the enemy, in order to ensure optimal preparation for the political conditions of the next phase. However, in a massive change from the past, no matter how necessary it is, this victory will only be a victory of a single stage, the first milestone in a long and arbitrary progression towards the political success of the intervention, for which it is only beginning to set in place the conditions.

This coercion phase, which should be as short as possible, is followed by what is usually a long transition phase –referred to as the stabilisation phase– during which military force continues to play a central role, while attempting to establish the conditions required to exit from the crisis. Acting on behalf of the population, and in close contact with it, the armed forces consolidate the temporary order put in place, by reducing and containing the level of violence as far as possible. Engaged in what could be called the principle of "positive de-escalation", they gradually restore the feeling of security, and assist in returning to normality and reviving economic and social life. It is no longer a question of destroying, but, with the help of a wide range of varied measures, of eradicating an often deep-rooted sub-stratum of violence within the society. During the course of this essential phase, the situation will often remain precarious, volatile, inconsistent across the zone and liable to sudden outbreaks of violence. At one and the same time, quite diverse military responses may be required, varying from forceful action to humanitarian relief. This phase is always very demanding in terms of numbers and actual presence on the terrain, since, while maintaining contact, it is necessary to carry out constant missions to control complex physical and human environment, while holding on to sufficient reserves to deal swiftly with any sparks that may threaten to engulf the theatre once more. The armed forces act in collaboration with the non-military actors, whose role is of ever-increasing importance. In contrast to the previous phase, the objectives set for the armed forces are rarely defined in detail, even if, overall, the intention is to restore stability by taking control of the area and re-establishing an atmosphere of trust between the protagonists.

At a given moment, if the stabilisation phase has been successful, it will give way to a third phase, intended to ensure a return to normality, in which the armed forces will take a back seat to the civilian institutions as they gradually return to their major role. This phase marks a return to peace. During it, the role of the armed forces is that of prevention and deterrence, in support of governmental and non-governmental institutions, both national and international. The gradual withdrawal of the armed forces, in favour of the legitimate authorities, local security forces and non-military actors, marks the definitive success of the military operation. In contrast to traditional military ideas, success here is not indicated by control, but by the lack of a need for control.

Success means not simply peace, but a self-sustaining peace.

STABILISATION – THE DECISIVE PHASE

Thus, in tomorrow's war, the timing of the decisive phase for the success of political action has changed. In the plan proposed, the central phase –during which the social contract will be restored– is crucial. It contains the most difficult conditions for military action and the main causes of a possible failure to restore peace. The decisive phase is no longer the short initial phase –the intervention– but, quite clearly, the longer consolidation phase which succeeds it. Recently lauded concepts, such as the "Rapid Decisive Operation" and "Shock and Awe", no longer have any relevance since, apart from the technical effect they produce, we now know that by themselves they are incapable of producing the desired political effect. What we have learnt, in particular from the operations in Afghanistan (2001) and Iraq (2003) is that, even if the coercion phase is extremely violent and utilises modern technological capability to its best effect, it is the stabilisation phase –the progressive return to normality through a sustained presence on the ground, in contact with the population– which, with the appropriate use of the force, is truly decisive. Thus, while it is essential for armed forces to have a high-tech coercive capacity, they should no longer be designed for the sole purpose of confronting their opponents, but also with a view to producing "effective security" and being able to restore civilian peace; in other words, with a view to reconstruction as much as destruction.

Stabilisation has two facets. Through its decisive nature, it is proving to be a valuable strategic function and important aspect of the external action. At a tactical level, it also the most essential phase of the intervention as a whole. It is essential, because it is during this phase that the political and strategic success or failure of the military intervention is determined. In the same way as we speak of a "major effect" when discussing tactics, we can speak here of a "major effort" or "major phase", since success in this phase "guarantees a successful outcome for the intervention", just as "achieving the major effect guarantees the success of the tactical manoeuvre". Speaking in the Palais Bourbon on 18 October 2006, the Chief of the French Army Staff made his views quite clear: "The notion of the decisive battle no longer has the same relevance as in the past. The main effort now focuses on the stabilisation of the situation on the ground." Being both an objective and a strategic effect, the stabilisation phase is truly the key phase, the decisive phase; in other words, the phase that determines the strategic result and facilitates the political effect.

Of course, as we have seen, this phase does not exist in a vacuum, since it follows the intervention phase which, by introducing an armed force, often in a violent manner, seeks to impose a temporary order, which must be exploited to the full. It is easy to understand that the success of the stabilisation is closely linked to the conditions in which this initial essential –yet preparatory– phase of the intervention has taken place. When planning the intervention phase, consideration must therefore be given to the stabilisation phase and how it is to be conducted. In its turn, the stabilisation phase is planned on the basis of the desired end-state, with each action being assessed less in the light of its immediate technical results, and more in respect of its ultimate political consequences.

The stabilisation phase is a complex phase because, rather than being homogenous, it is fundamentally hybrid, comprising moments of coercion and taking control of violence, as well as moments of high and of low intensity, as described by the Chief of the French Army Staff, speaking on 19 October 2006, at the Ecole Militaire: "Stabilisation does not mark the beginning of a period of peace; on the contrary, it is a period of confrontation in all areas, civilian and military. For this reason, we must be careful to avoid thinking in terms of a linear evolution of the situation. The storm may follow the calm with no advance warning." It is also hybrid in that success can only be achieved through the convergence of a number of diverse "lines of action" –of which the "conventional" military line is only one component– and through the complementary nature of joint, and in particular inter-ministerial, actions. Again in the words of General Cuche: "Only the manoeuvre, in other words the combination of military and civilian means, is capable of producing the effects which allow armed groups to be neutralised or discredited and which offer favourable prospects to the population groups."[1]

In this decisive phase, the land forces –with their ability to destroy, protect and to build, at the same time and with the same unit– play an essential role. This they must do without losing sight of the political purpose which outweighs the technical capability. This is a demanding requirement. As Gallieni and Lyautey have already emphasised, one does not approach a village in the same way if it is to be destroyed as an obstacle on a strategic penetration route into the combat zone, or if the intention is to gradually establish the conditions for peace in a crisis theatre. In determining his actions and his modes of war, the commander must always bear in mind the "normalisation" phase and the desired end-state. The challenge facing the armed forces, and the army in particular, is thus exceedingly complex. On the one hand, it is necessary to retain a strong capability, in order to be able to make a clear impact in an increasingly short dissymmetric phase and during specific coercion operations while, on the other hand, being able on a daily basis to conduct actions often wrongly described as "low intensity", but fundamental to the ultimate success of the intervention.

---

1.   18 October 2006, National defence and armed forces committee.

CHAPTER VI

A NEW PURPOSE FOR MILITARY ACTION

The continuum described above indicates that, to a certain extent, we are returning to the classic view of the purpose of military action. In the recent past, we believed that military success led directly to the strategic goal. This idea is dead.

Now, we see that military action no longer necessarily pursues the absolute and concrete objectives of the industrial war; rather, it pursues more malleable goals which emerge from the dynamics of the action.

More importantly, we are coming to the realisation that military success simply leads to establishing the conditions required for strategic success.

A CHANGED PURPOSE

In contrast to the idea that lay behind the potential cataclysmic confrontation of the Cold War, war is a continuous process. It strives to open up, maintain and consolidate a space for negotiation and diplomacy. In other words, in true Clausewitzian tradition, which is at odds with American strategic culture, it does not represent a break, but rather one stage in a political process. The political dimension exists before the war, it continues during the war through the decision to engage armed forces, and it carries on after the war; there is no interruption of this course. "War does not cause political interaction to cease, nor does it change it into something different; it continues to exist, whatever may be the form of the means which it uses. The chief lines on which the events of the war progress, and to which they are attached, are only the general features of policy which run all through the war until peace is achieved."[1] Clausewitz's observation is as relevant today as it was then.

_____
1. Clausewitz, p. 605.

It will only be possible to judge tomorrow's war in the light of the political results to which it has contributed.

Thus, while the initial military objective can be achieved by means of military force, this does not apply to the political result, which can only be the later consequence of such military success. The final "victory" –though this is no longer the most appropriate word since the concept of victory belongs to strategy rather than politics, and tomorrow's wars will be fundamentally political– is not a military result; the evolution of the social, economic and political context will be the most important aspect. It is essential to understand fully this evolution and all its consequences, the most important of which is that the objective of military strategy cannot be the same as that of the political approach; the former must lead to the latter, but cannot be assimilated into it.

To use today's buzzwords, the military instrument is no longer an "effector", but has now become a "mediator", mediating between political will and political success. It is not a question of "winning the war"; "winning the battle" becomes an intermediate objective – essential, but still intermediate. The French Chief of the Defence Staff put it quite clearly: "It is not that military intervention is of no use; it is just that, by itself, it is not enough to resolve a conflict."[1] Thus, for the armed forces, it is not enough to "win the war", nor to "win the peace"; their role is to create the conditions that will allow peace to be established. Questioned in July 2007 about the conditions for success in Iraq, General Petraeus, commander-in-chief of the Coalition forces stated: "Success in Iraq will be defined not in military terms, but in political terms. Military action is necessary, but is not enough to ensure long-term political stability."[2]

There is no doubt that Western forces produce their best results in difficult phases during times of crisis, in the operational phases of emergency operations, the phases of military action relating to the de-escalation of violence and the immediate phases of humanitarian action. This is probably because these phases correspond to the Western sense of the immediate. However, they still have plenty of room for improvement with regard to longer term actions, such as the gradual resolution of crises, probably because this longer horizon does not fit in with our preferred mode of strategic thinking, or perhaps because our skills in dealing with short-term technical matters render us less able to grasp the subtleties of the political longer term. In the same way, while we in the West perform excellently in

---

1. Jean-Louis Georgelin, in *L'armée française face aux nouvelles menaces* [The French Army and the New Threats], *Politique Internationale* [International policy], no. 116, Summer 2007, p. 152.
2. In *Valeurs actuelles* [Current values], no. 3688, 3 – 8 August 2007.

phases with a negative objective (such as ending an attack, a massacre or a famine), we still have much to learn when the desired objective is positive (rebuilding the social framework, living conditions, the minimum state, etc.).

Changes in the context of crises, and thus the wars we are likely to encounter, mean that we must make an effort to improve our transition from a military sphere to a security sphere, or from urgent humanitarian assistance to the politics of reconstruction and development. They also mean that the new limits of military effectiveness must be explained unremittingly to the politicians responsible for the decision to intervene militarily. In addition, we must demonstrate the importance of an overall view of the action, with its various civilian and military components, whereby the objective cannot be purely military. They must constantly be reminded of the practical and political conditions of the utility of the force. If not, the coffers of the state will have been emptied to no effect and the blood of men spilt in vain, with the only result that the West has become a little less credible. Military intervention must not be allowed to have as its sole aim the need to "do something" or to salve consciences.

EXPANDING THE RANGE OF MILITARY MISSIONS

One of the most important consequences of this change to the purpose of military missions is an expansion of the range of military missions. The notion of "mission creep" can thus be replaced by that of "implied tasks".

This is true for the positive aspects, but equally so for the reverse.

From a positive point of view, the military mission is to "create the conditions needed to achieve the strategic effect". Such conditions always involve establishing normal living conditions for the population, which itself has now become the goal: security conditions, humanitarian conditions, etc.

This also applies in the converse situation since, in fact, the armed force has no choice other than to protect and support the population. Others could perhaps do it as well as or even better than the military, but they will not be there. The military will always be used to fulfil this role for the time needed to re-establish a sufficient level of security and stability to allow civilian agencies to take over this work, for however long this may take. Experience shows us, that for some considerable time, only military manpower will be available to carry out certain essential tasks, particularly in the absence of a local civil authority, which is usually the case. It is simply a question of available capacity in a given situation. In the initial stages at least, there are no inter-agency structures equivalent to a military head-quarters. The civilian component of an operation does not have the organisation, the staff, the training, the planning tools or the command and control resources to deal with all the assets made available to carry out state building and defuse potential insurrections, which may be either vertical (against the established power) or horizontal (between groups and factions). It is essential to have the conceptual and physical means to carry out these tasks satisfactorily, as there is no second chance. Success starts straight away or it does not start at all; it is the problem of the "Golden Hour", beloved of emergency doctors, which has to be planned and prepared.

The military are very closely involved in each step of the success or failure. They are thus unable to disregard these imperatives, even if they lie outside their so-called conventional expertise. These civilian tasks become military tasks by default, as soon as the heavy weapons fall silent. If the force does not fulfil these roles, previous military successes will have been in vain, thereby removing their legitimacy. The force must be capable of carrying out these missions.

On the other hand, the force must not allow itself to get bogged down in projects which go well beyond its capabilities and its competence: it is not easy to strike the right balance. The military command must be clear about what it is willing to do –because it is strictly essential– and what it will not do. This "farewell to arms" on the part of violent factions and actors can only be "prepared" by the military forces. It is a question of capability and of the sound management of the nation's resources. The baton must be passed to other capable actors, who, with the limited and decreasing assistance of the military force, are able to complete the third phase, the return to normality.

The example of US General Chiarelli, credited with a major success while commanding the 1st (US) Cavalry Division in 2004-2005, when he was responsible for the zone around Baghdad, is very vivid. Before his unit deployed to Iraq, he sent his staff officers to work with the municipality of Houston, believing that it was just as necessary for them to understand how the city functions, as to know how to capture terrorists. In his eyes, the only the way of reducing the insurgency was "to shoot less and build more".

It is clear that the evolution in the conflict situation has led, with the very simple aim of improving technical effectiveness, to a new requirement being placed on the forces: that of supporting the population. This goes far beyond the concept, previously thought sufficient, of civil-military action carried out solely with the aim of achieving acceptance for the force.

Land forces –which now carry out the vast majority of engagements– will naturally be involved in tasks which, under normal circumstances, would be the province of civilian organisations. They need to prepare for this in terms of doctrine, equipment and instruction, as well as in terms of officer and NCO training, which must be expanded to include civilian disciplines required for the reconstruction of a country in the widest sense. It is, however, extremely important to remember that it will only be possible to carry out these missions effectively if, at all times, the force has clear military superiority over its potential opponents.

It is interesting to note here, that this "need to protect" stems from a changed and fairly recent understanding of effective crisis resolution. In contrast, history offers many examples of the exploitation of civilian populations, such as General Sherman deliberately terrorising the civilian populations of Georgia and the Carolinas during the Civil War or the strategic bombing of Guernica, Coventry, Dresden, Hamburg, Tokyo, Nagasaki or Hanoi. This use of air power had previously been put forward –first by Giulio Douhet and later by Mitchell– as a means of affecting the opposing political power through the suffering of its citizens, which is not far removed from the concept of using strategic nuclear weapons mentioned earlier.

It would, of course, be wrong to claim the "fungibility of power"[1] and to assume that a military capacity could be easily converted to a social or political capacity. However, it is not enough to disregard reality and to say that providing protection and assistance to the population are not a matter for the military. They are: they constitute the effectiveness and ultimate purpose of the armed force.

---

1. Expression used by Bertrand Badie, *Les Champs de Mars no. 17, La documentation française, 2005, p. 7.*

PART TWO

THE NEW CONDITIONS
FOR MILITARY EFFECTIVENESS

"The world we have created is a product of our thinking; the problems it causes cannot be changed without changing our thinking."

Einstein

Chapter I

GENERAL CONDITIONS

In the past, there was a force model, consolidated by the succession of centuries of Westphalian thinking and further reinforced by the characteristics of the confrontation between blocs, which corresponded to the state-oriented uniformity of the threat.

While there is no doubt that we must retain the capacity to conduct conventional wars in order to reduce the chances of them occurring, and to produce, during interventions, a level of violence that will constrain the opponent, there is nonetheless a need to adapt the defence capabilities. A world characterised by the coexistence of a number of new conflict perspectives needs the corresponding revised force models, able to ensure both a full understanding of the new conditions of these engagements and their objective.

The experience of the last fifteen years enables us to identify two major principles for action: modesty and balance.

Modesty

First of all, modesty, since the failure of Western intervention stems all too often from overly ambitious goals, in terms of security, social conduct, governance, education, etc., that is to say, from an incorrect initial analysis. In particular, the notion of normality needs to be re-examined in this context. It is pointless and presumptuous to attempt to impose "our normality" on the countries in which we are operating, when "their normality" may still include, for example, a certain degree of authoritarianism, corruption, inequality, organised crime etc. The return to what was normality before the crisis is a reasonable initial objective; the road to our normality may often take several decades. Our idealism should not allow us to believe that our model is universally envied, nor should we forget that the strength of traditional social and cultural models means that local populations are unlikely to adopt immediately political models that the West has constructed over the course of a prolonged and painful period.

As a general rule, it is therefore better to adapt existing institutions rather than to impose major changes on them. The ultimate aim is a country at peace with itself and its neighbours. Democratisation should be seen as one possible means rather than an end. Dictatorial power is the product of a society and its system; it will therefore take much more than the toppling of the dictator –and of his statue!– to achieve any deep-seated changes. Furthermore, and to deal once and for all with a dangerous expression, while it is possible to reconstruct a minimum state, in accordance with locally and psychologically acceptable norms, it is without doubt unrealistic to attempt to "build a nation": we are fully aware that a task of this magnitude requires many centuries…

Balance

Secondly, balance, because the second most common cause of failure is the discrepancy between the means deployed –the military means, but also those relating to the willingness of the states involved to make a long-term investment– and the objectives sought, bearing in mind the local conditions. The size of the country and of the population are important, but the strength of the probable opponent is crucial since our, often painful, experience shows that if the ratio is less than 15 or 20 to 1^1, a counter-insurgency force has almost no chance of winning. Success will also be impossible if the assets deployed are insufficient to respond to attacks. In Iraq, one of the immediate and lasting problems encountered has been the disparity between what the population was hoping for, in terms of reconstruction after three decades of negligence, and what the coalition was capable of achieving, and of achieving rapidly, in order to prevent disappointment. It is in our short and long-term interests that promises be kept. It is therefore crucial not to promise what we cannot deliver, and to avoid creating false hopes. Here we become aware of the limitations and weaknesses of our impatient democracies; success is a long-term process, but we live under the constraints of requiring a rapid return on investment. The pace of military effectiveness in tomorrow's war and the pace of political life in our capitals are very different.

Indeed, this question of balance is crucial. The legitimacy of the intervention rests ultimately on the existence of a balance; if the chances of achieving a better state are non-existent, the intervention cannot be justified, no matter how indisputable the soundness of the course or how politically forceful the compassion of the television audience. In particular, as contemporary operations have demonstrated the inability of force alone to produce the desired political results, it is preferable not to intervene in cases where it is unlikely that the other resources required for the action can be made available.

---

1. This is the same as the figure currently considered to apply to combat in urban areas.

Cohesion, an additional dimension of such balance is also indispensable: that is to say, cohesion between the aims, the ways and the means, during both planning and execution, but primarily cohesion in order to ensure the overall effectiveness of the actions of the various actors involved. It implies the priority of political aspects and the unique nature of the objective. It also implies unified overall management, encompassing the various lines of operation. Finally, it implies the integration of actions. Such cohesion is established before the action, is managed during the action and offers a guarantee of the sustainable achievement of the desired end state.

AN UNAVOIDABLE COROLLARY

The effectiveness of military action is dependent on its acceptability – for the "intervener" and, as far as possible, also for the "intervenee". Applying principles whose universal nature is apparent only to those who are keen followers is unlikely to be enough to act as a basis for acceptability. Thus the concept of a "just war" is a dangerous one, able to legitimise neither the compassion nor the idealism so widespread today.

To regain its legitimacy, war –these days frequently edged out of the political field– must return to its initial purpose, which was precisely a political one. The moral attitude of states is naturally ambiguous since it is a reflection of power. It fluctuates because states have, first and foremost, duties and interests. Much like that of individuals and other social groups, their ethical code is shaped by the need to face the rest of the world: they tolerate what is required for their security. Paradoxically, the return of moral considerations to international relations weakens the Westphalian system and tends to weaken the notion of state and sovereignty, that most interventions are aimed at strengthening. Although one can only commend the moralisation of foreign policy, in other words the development of a policy including moral considerations, war and military force must remain fundamentally amoral tools.

While the idea of a "justified war" is admissible, that of a "just war", laden as it is with moral pretension, is not. It removes the moral culpability related to the use of violence and leaves the door open to excess, through the total nature of its objectives. Morals reject compromise and encourage immoderate behaviour; the use of violence in support of the good –be it the spread of Christianity, the imposition of revolutionary communism or the triumph of democracy– is much more difficult to moderate than violence in support of law. Thus, while the first Gulf War was a legitimate and limited war, an instrument of policy, the second claimed to be a "just" war and a moral instrument. It is now becoming a total war, with no compromise possible between the parties and no foreseeable end, and with dramatic consequences that are spreading throughout the world.

History also demonstrates that only the certainty of impunity allows the action to be based on moral considerations. Only absolute power can have the luxury of defining what is "good", which constitutes the right of force. Monique Canto-Sperber observes that: "In the tradition of the just war, the precise and detailed analysis of the causes and means of the war was intended to determine its fairness and the moral status of the protagonist. Today, it is this moral status which partly decides whether the conflict is just. Certain states, which consider themselves to be moral states, use this fact as a basis for justifying the morality of the war they are conducting."[1] Nowadays, the difficulties confronting Western power make it impossible to develop the policy of "its moral code": gone are the days when Western countries could intervene anywhere they considered it necessary solely on the basis of their virtues.

The concept of a "just war" implicitly assumes inequality between the opponents –in particular, moral dissymmetry and an imbalance of rights– which is the cause of severe excess and deep-seated hate. In this vision, the opponent "of whom it has been decreed that he must be submitted to judgement, and liable to receive a punishment suited to his crime, passes into the jurisdiction of the righter of wrongs."[2] The behaviour of the American guards in the Iraqi Abu Ghraib prison in 2004 probably illustrates certain excesses of such inequality. This moral dissymmetry is not, however, a factor of effectiveness. On an individual level, it will make more difficult the contacts with the opponent that will be essential once the guns have fallen silent in tomorrow's war. From general to private, it is more difficult to establish a dialogue with someone who has officially carried the weapons of "Evil", than with someone who, during the confrontation, was technically only an opponent. This recalls the episode when General Eisenhower refused to meet and receive the sword of a German general whose army he had just defeated, because it was not right to set up a dialogue with the Evil opponent. Today's opponent must be respected, because he will frequently be a partner in seeking compromise solutions, rebuilding and governing. He may be tomorrow's ally. François Mitterrand pointed out this fact when speaking to the Knesset in Israel: "Peace must be made with those against whom you are fighting."[3]

---

1. Canto-Sperber, p. 289.
2. Frédéric Gros, p. 184.
3. In *Continuer l'histoire* [Continuing history], Hubert Védrine, Fayard, Paris, 2007, p. 144.

In a "justified war" –for humanitarian reasons, for example– the military force preserves its value of neutral morality. The force acts in an attempt to achieve the effect requested of it, but does not take sides.

BEYOND DESTRUCTION AS A MAJOR ARGUMENT FOR MILITARY EFFECTIVENESS

The quantitative approach which lies at the direct origin of French military thinking –as well as indirectly, through the strong influence of American military culture– tends to favour the act of destruction and to give little account to other, less weighty dimensions. However, these dimensions are now proving essential. The current evolution in the conditions of use of forces reduces the relevance of this approach. The disappearance of states from war has decreased the destructive capability which, effectively or potentially, has traditionally been the central mode of action for armed forces confronting their peers.

Destruction has now reached the limits of its effectiveness since, in tomorrow's war, political factors will dominate. Technological superiority no longer deters the non-state opponent. The notion of capabilities, confronting the capacities of the two sides, does not impress him. The evolution of conflicts –in Iraq and in Afghanistan, in particular– indicates that conventional strength has lost a great deal of its deterrent effect on the potential opponent: why should someone who has already opted for certain death worry about the possibility of being killed? The new actors are no longer seeking superiority since they know that a military victory –in the conventional sense and with conventional means– is no longer a prerequisite for a political victory. In such circumstances, a stand-off "fire and forget" strike is often pointless. In terms of ultimate effectiveness, it is overtaken by the principle of contact action on a continuous basis, which makes it easier to achieve the essential outcomes resulting from the factors of influence and psychology.

Taking the easy solution of employing only traditional means to achieve the material destruction of the non-conventional opponent has never produced the desired political result. Once it is no longer a case of reducing a state's ability to resist the will of another party, the means intended to destroy states' confrontational capability are of no further use. This is particularly true since, by their very nature, non-state assets are much less vulnerable to conventional strikes, presenting virtually no targets for high-technology firepower.

Other than the possible and brief advantage it may offer, destruction may also prove decidedly counter-productive. From a tactical point of view, it is by no means certain that destroying the enemy's "system", if one exists, is truly advantageous. The collapse of this system in itself creates destruction on the battlefield, causing greater uncertainty by shattering the opponent into autonomous combat cells. Moreover, attacking the "enemy system", as discussed by Warden in particular, leads immediately to the "totalisation" of the conflict by decompartmentalising the use of violence. Extreme levels are reached instantaneously since, at the same time, the air campaign will affect not only military targets, but also the political environment and infrastructure; in other words, the population, who will thus be "totally" involved in the war... something that will not easily be forgotten. Attacking vital infrastructure or needlessly weakening governmental structures, which will almost certainly be needed to provide support during the stabilisation phase, is not necessarily the best recipe for the political success of the intervention. In the past, the consequences of technological progress –in particular the new capability for stand-off precision firing– seemed to indicate that it would be possible to reduce the political cost of interventions and thus, in parallel, to increase political and operational freedom of action. Unfortunately, although the impression was not entirely false, interventions over the last ten years have shown that low political costs have primarily been linked to poor political results.

Compared to the bad ways of the past, we are now observing a reversal in the process of achieving military effectiveness. Whatever the methods invented by those thinking in terms of time and technology, in the past, for Western forces, the fundamental principle was to destroy the opposing force in order to break its political will. In contrast, the current principle, using a moderate degree of force ourselves, is to influence the will of the opponent so that he rejects the use of force. This led Sir Rupert Smith to observe, rightly, that armed confrontations are now more "a clash of wills" and less of a "trial of strength", in the words of Clausewitz. Thus today, traditional strategies and weaponry are losing their usefulness. This shift is all the more evident as, because the new threats are no longer purely military, it is precisely the ability to produce non-military effects that is becoming a mark of utility for modern forces. If we are serious in our desire to increase the reign of democracy across the world, we should follow the advice of that great American diplomat, George Kennan: "It is time to recognise that the true aims of a democratic society cannot be achieved by war and mass destruction."[1]

---

1. Extract from a speech given by George Kennan at the International Atomic Energy Council in 1949, quoted in Strategies of Containment: Critical appraisal of Post-war American National Security Policy, Oxford University Press, 1982.

In parallel with the gradual decline of destruction as a political tool, we see that it is also losing its legitimacy. This change results from two phenomena.

The first is, paradoxically, the information revolution. Nowadays, any act of destruction –and its inevitable collateral damage– is presented rapidly for trial by international public opinion, which is swift to judge according to criteria far removed from political principles; advanced societies are very sensitive to the round-the-clock portrayal of violence on their screens. The second phenomenon is that countries frequently engage their forces for interests which are no longer vital or at least do not appear so to their citizens. In this way, while the vital nature of the interest warranted the action of destruction, in the spirit of legitimate defence, the relative nature of the interest being defended no longer justifies it to the international conscience. Destruction reduces the legitimacy which is seen today as a condition of freedom of action.

However, if destruction alone is now less able to lead to political solutions to crises in which the use of the armed forces is justified, it nevertheless retains a role of "reassurance". There is no doubt that, faced with a new enemy who may be erratic and has limited modern capabilities, the specific destructive capacity of air weapons offers the ultimate guarantee for the troops in contact combat. This phenomenon is particularly evident in Afghanistan today, where close air support is an indispensable aspect of effectiveness in tactical combat against the Taliban. The army of a major power with a true joint capability cannot be compared to an army which does not have this backup, even if the armies themselves are comparable. In the same way, since simply projecting power is usually unsuccessful in achieving a political result, the projection of a contact army –in other words, the indispensable projection of forces– demands a multiform joint capability.

THE FRAMEWORK CRITERIA

Changes in societies and the contexts in which forces are deployed, combined with the immediate and universal availability of information, have reinforced three essential criteria for the acceptability of military violence, and thus for political freedom of action.

The first is the respect of the essential values of both the intervening nations and those subject to the intervention.[1] For the former, these values

---

1. The second of the ten directives issued to these units in the mid-2007 by General Petraeus, commander of the Coalition forces in Iraq, reflects this: "Give the people justice and honour." Since justice and honour are key elements of the Iraqi culture, it is important to treat the people with respect and dignity, and also to act openly to combat any form of injustice.

constitute the foundation of both "law in war" and the "law of war". It is worth noting that this "right to violence", the basis of the decision to use force, is by no means absolute: it is relative and changeable. It is dependent not on "objective" moral values, but rather on the perception of the threat and the sense of relative power –and thus of impunity– felt by the intervening nation. This sometimes causes the country in question to exaggerate the risk felt by its own national public opinion, in order to increase its moral right to external violence. The need to respect the nation or nations subjected to the intervention makes it necessary to assess carefully the "normality" that can be restored; Western normality would frequently be seen as "abnormal" there. To attempt to impose it harshly over a number of years is clearly presumptuous, as we have said, but would also be counter-productive in terms of the objectives behind the true reasons for the intervention.

The second criterion is legality. The wide range of legal instruments and special international tribunals means that military action must take place in a clear legal context. The problem here is the absence of universally recognised positive international law; in the absence of such a legal framework, it is the UN, whose decisions are considered to represent the majority feeling, and thus the majority force, which lays down the law and the norms.

The third criterion, and enhanced synthesis of the first two, is legitimacy: it presupposes legality but goes much further. Its key characteristic is that it is now no longer either accorded or acquired definitively.

The first difficulty encountered by the intervening power is that, by its very nature, legitimacy is uncertain, since it can only be discerned once the action has been completed, even if the legal conditions prevailing at the start seemed to justify it. At the end of the action, legitimacy is dependent on the concrete results of the intervention. But, even if legitimacy has been accorded later and by others, it is important to take account of the way in which it is perceived from outside and to understand its impact on the success or failure of post-intervention management. Monique Canto-Sperber writes quite correctly that: "It is only after the war has finished, in a post-war environment that is difficult to define, that we will be able to say whether the war was just. The moral shakiness is a given. War will in part be judged in the future perfect."[1] After all, and on these same lines, Nietzsche was not completely wrong when he wrote: "It is the good war that hallows every cause."

---

1. Canto-Sperber, p. 304.

Thus, legitimacy cannot be decreed. It must be established within a perception and international public opinion is not shaped by the laws of war. Being very closely linked to the conduct of an intervention from the outset, legitimacy must be viewed within a dynamic context. In other words, it must be established, consolidated and preserved at all times, within the fragile atmosphere of a perception. The fact that it is bound to the choice of ways and means, to the methods and the intensity of the force and to the rules of engagement, renders legitimacy unstable, depending every day on how the local population and international public opinion perceive events and actions. In a way, the manners of war, the application of *jus in bello*, determine retrospectively the *jus ad bellum*. The more fragile the legitimacy, the less the force can afford an error and the more it is constrained in its freedom of action: "The best of reasons for fighting a war can, at a later stage, be disqualified by the course of that war."[1]

In need of permanent protection and consolidation, legitimacy also depends heavily of communication strategies, even where it seems to be self-evident since, in contemporary thinking, a justified war and a legal war often coincide only for short periods.

Looking beyond legitimacy, we encounter the idea of strict necessity or strict proportionality. In addition to the need for sound judgement with regard to the options for employing the armed force, this idea demands the constant ability to modulate and adjust the force and to reverse the modes and means of action, in order to have the clear image of the environment needed when in contact. Rather than "a minimum force", we should refer to "the minimum force needed", since it must be determined as a function of the ultimate purpose. This proportionality relates not only to specific incidents where force is used, but also to the intervention as a whole. In the words of St Thomas Aquinas, which are as true today as they were in the past, an intervention is only legitimate if the reasons for intervening are proportional to the wrong endured, and to the possible good that may come from it. In the eyes of international public opinion, it is precisely this absence of proportion which seems to have been missing from the conduct of engagements, particular the air engagements, by the Israelis in the summer of 2006.

Proportionality must be accompanied by discrimination. Collateral damage has become less and less acceptable; it weakens the legitimacy of the intervention in the eyes of both local and international opinion. This negative effect can be felt particularly strongly in operations conducted against terrorists, because it allows a parallel to be drawn between the

---

1. Canto-Sperber, p. 343.

intervention and the terrorist act, the key characteristic of which is the lack of discrimination between victims.

SUSTAINED ACTION

Tomorrow's war will thus be as much about psychological areas as material areas, thereby implying a prolonged action. There will be no clear and military victory marking the end – indeed there will be no clear end. The nature of such a war will make it impossible for it to be contained in time. Such a limitation in time would assume a dialogue between interlocutors, but the removal of the state element and the dispersion, or even multiplicity, of an opponent who in many cases has no political unity make this impossible. The decline of states is now frequently at the origin of crises, and also of the problems in resolving them.

Even if the dynamics of the action has to be organised in such a way as to avoid getting bogged down or a feeling of stagnation, the pursuit of an immediate, or even rapid, result usually constitutes an error in the settlement of disputes born of a long history, which need to be dealt with in the same timeframe. A precipitate response will never produce stabilisation. Although it may be possible to deal rapidly with the symptoms, the same does not apply to the causes. Social evolution is a long-term process: success in dealing with a crisis is judged many years after the start of the intervention and not on the effectiveness of the initial action. The reconstruction of states and social contracts takes many decades[1], the reconstruction of nations many centuries.

The very unrealistic notion of "first in, first out"

The erroneously attractive notion –regularly rebutted by experience– of "first in, first out", in other words the notion of a precisely targeted intervention followed by a rapid withdrawal before the political resolution

---

1. In this context, the example of the conflict in Northern Ireland is particularly interesting. Despite operating among a population which shared their culture, on their own territory, as a national force and within a simple legal framework (their own), it took the British 38 years to resolve "their" Irish problem. (From the bombings in 1969 and the deployment of the Army in Northern Ireland, to the "farewell to arms" of the IRA in July 2005, and then to the recognition by Sinn Fein in January 2007 of the legitimacy of the Northern Irish police and judicial system, to the final departure of the British troops, who officially left on 1 August 2007, leaving the police in full charge.) This consisted of around four years to stabilise the conflict –which produced neither winner, nor loser– and 34 more years (and 700 dead, more than in all their other operations –the Falklands, the Gulf Wars, Afghanistan etc.– over the same period) to arrive at a political solution acceptable to both parties.

of the conflict, is not compatible with the new conflictual context, the current subordination of military imperatives to political decisions or the indispensable role in crisis resolution reserved solely for the armed forces. Having arrived here from across the Atlantic, the concept of FIFO has become entrenched in our staffs, always keen to seize on any overseas ideas that may be suitable for adapting to the reality of our scarce resources. It is true that, nowadays, rather than trying to hold on to the territory over which we have established control, through our military strength, there is a tendency to look for the quickest possible escape route. This said, clear-headed observers will have understood that the idea of a FIFO intervention amounts to a pure denial of reality.

The first aspect of this reality is that both the beginning and the end of intervention are decided at the political level; the military, quite rightly, simply have to conform. Thus, if the armed forces adjust their means and methods on the basis of the false assumption that it will be possible to withdraw whenever they want, they are certain to find themselves in difficult situations. The second mistake is to assume that Western troops, skilled in initial operations aimed at a high-technology military victory, would easily find back-up troops for the thankless task of obtaining, under difficult circumstances, the desired political result. This idea died –unequivocally– in the deserts of Iraq and the mountains of Afghanistan. Contrary to what was envisaged in the euphoria of the "unipolar moment", "superpowers sometimes need to clean windows and to accompany children to school". It is through these capabilities that nations establish their weight and their right to speak, since what really counts is not the initial coercive capability, but the ability to be involved in the crisis on a long-term basis, until the political result is obtained.

Tomorrow's war will still be a war and we will have no control over its duration.

The third, and strongest, aspect of reality relates to the persistent nature of war. It is circumstances that are in command rather than politicians; once you get involved you have no idea when you will be able to withdraw. Time has taken nothing away from the words of Clausewitz, who said: "Once politics has implemented it, war of its own volition usurps the role of politics; it relegates politics to the sidelines and regulates the event according to the laws of its own nature."[1] The weight of men, the weight of events –their sway, their dynamics– the force of operational logic, and the man of action who, "following his natural bent, pursues the action with no limits other than full accomplishment according to his own

---

1. Clausewitz, p. 87.

criteria"[1], sadly always triumph over theoretical planning. The departure of the force in fact no longer depends on either the soldier or the politician; once unleashed, the violence of war tends to follow its own paths. Far from being an inanimate element, it is possessed of a living force with unexpected and complex consequences, often very different from the aims for which it was initially deployed. Thus we see that war and its consequences, whatever their horizons, can never be predicted, control or fully defined. The comments made by General Dan Halutz[2] (Chief of Staff of the Israeli Defence Forces who, following the capture of two Israeli soldiers by Hezbollah, launched so-called "restricted" punitive bombing actions) on this subject are, once again, very pertinent: "That evening, 12 July, we did not know that we would be at war with Hezbollah." He continues: "The aims of the war changed with the expectations of public opinion." What followed is history... Very rapidly the operation exchanged its initial name of "Suitable Punishment" for a new name which spoke volumes, "Change of Direction"!

Future wars will be long, partly because achieving political success will be a long-term process and partly because future wars will still be war, in other words, a necessary yet dangerous instrument, requiring careful handling. Its "natural tendency" or its "strict logic", in the words of Clausewitz, mean that it is a capricious tool, awkward to use and with arbitrary results. Its use can only be a last resort for political action, with preference always being given to all preventive manoeuvres. Real war is not war as an "object", as it might be viewed independently of circumstances; real war is a "subject", with its own life and with its own objectives, that will eventually have a retrospective impact on the initial political objectives during the inevitable transformation of the "Ziel" and the "Zweck".[3] It is interesting here to read the words of a man very familiar with war, having been involved in it and having conducted it at the highest level – Winston Churchill. This old hand at both politics and war reminds us of the need to learn from the past. "Never, never, never believe any war will be smooth and easy, or that anyone who embarks on the strange voyage can measure the tides and hurricanes he will encounter. The statesman who yields to war fever must realise that once the signal is given, he is no longer the master of policy, but the slave of unforeseeable and uncontrollable events."[4]

---

1. Delmas, p. 244.
2. Statements made on 20 October 2006.
3. The "goal of war" and the "purpose of war".
4. Winston Churchill, p. 246.

The inescapable truth is that it is always more difficult to extract one-self from a war than to join it, and that devising exit strategies on the basis of an inflexible agenda is as unrealistic as it is counter-productive. Recon-struction is incompatible with restricted planning which allows the oppo-nent to organise his own actions; the schedule for exiting from the crisis must remain very flexible. In other words, to orientate our actions towards an "end date", rather than an "end state" is a serious error. Let us be in no doubt, the opponent and the circumstances will always be key factors in future wars: the "fog of peace" is just as thick as the "fog of war".

We should be aware that a "quick fix" solution, no matter how attrac-tive it might seem, is really of no use. We need to prepare ourselves for long operations. If we are lucky we may be pleasantly surprised.

Speed is no longer the key to success

The need for a sustained approach to solving crises calls into question a whole way of thinking, which is based on the idea that military victory lies primarily in the ability to predict and react more quickly than the opponent. This theory underlies the American Transformation[1] move-ment. Transformation, based on increasing the pace, retains part of its tac-tical relevance during the initial brief intervention phases, but is now less universally relevant because crises and conflicts are resolved within popu-lations, who live at a different rhythm. Within societies where the times-cale differs from our own, the time needed to resolve a crisis cannot be the virtual speed of OODA[2], or the brief periods of Western impatience; it can only be the longer horizon of changing reality. The "real time" of plasma screens bears very little relation to the "actual time" of crisis resolution. The speed of technical success may well have become the priority of mili-tary instruments, but it in no way guarantees political effectiveness. We have designed the structures of our forces to suit the requirements of our Western political rhythms, which demand rapid results, but the political timepieces of the countries in which we intervene are totally out of sync with ours. General Petraeus, commander of the Coalition forces in Iraq said it himself in June 2007: "The Washington clock is running a lot faster than the Baghdad clock."

---

1. The subject of Transformation is dealt with at greater length in the next chapter.
2. Observation, Orientation, Decision, Action. Devised by UASF Colonel John Boyd, the concept of OODA is in current usage to describe the decision-making cycle. In conven-tional combat, the winner is the side that has a faster OODA loop than his opponent, making the best possible use of the revolution in information technology.

However, it goes further than this. This speed –which had become the essential dream and which only yesterday was considered the key factor for victory– now seems in most cases to be counter-productive as it makes impossible to comprehend the situation fully. This makes it impossible to adapt and, in the eyes of the populations, represents the very essence of the lack of willingness to make a commitment. Tactically speaking, in tomorrow's war of counter-insurgency, the rapid destruction of a newly acquired target may even prove to be the least sensible option.

In addition, the desired increase in the speed of operations and political actions naturally results in an increased risk of incorrect interpretations and decisions, thereby promoting the increased use of technology by the armed forces, considered, wrongly, to be a means of avoiding errors. Having analysed current transformational trends, Joseph Henrotin refers rightly to: "The emergence of a chronostrategy, [which] being highly techno-centric, also has a direct influence on political positions, both at international level and in defence policy and national security strategy." This premium accorded to speed affects the conduct of external policies, favouring the use of the most offensive options over options able to produce deep-seated effects, but which require more time.

An inevitably heavy commitment

In tomorrow's war it will be more a case of long-term persuasion, through contact and dialogue, than short-term imposition, through constraint and the capacity to destroy. What accompanies the contact will be as important as the intervention. In this sense the trend (born of the technological advances of the Cold War) of removing the soldier from the battlefield does not favour sustainable solutions. To avoid failure, the intervening countries must make a resolute long-term commitment (security, governance, economy, etc.). The population will only trust the intervening nation, and participate in the proposed reconstruction process, if it knows that the coalition will remain for the time required and will not abandon it to its old demons and unforgiving retribution. Success depends on establishing a credible alternative vision of a more attractive future than that offered by the irregular forces, thus on the slow and gradual reshaping of attitudes. Attitudes can be neither formed nor changed overnight.

Of course, the tools needed to effect this slow transformation are not only military tools; on the contrary, their gradual disappearance is the true mark of success for an intervention in which they were initially the spearhead. When deciding to launch the intervention, the participating nations must be fully aware of this obligation. They should understand that the legitimacy of their intervention will ultimately rest on their willingness to commit to significant investments over a prolonged period of time. This is a commitment which involves an obligation. A recent official study carried out by the Pentagon shows that, between 1990 and 2006, the loss of human life during stabilisation operations was six times as great as during conventional operations, and that, in the same conditions, the ratio of cost[1] was 5 to 1. Guidance and follow-up, however costly, are a prerequisite for the utility of the operation. Without them, the risk of a "relapse" is very high. The World Bank has come to the alarming conclusion that, during the first five years of peace, countries emerging from a period of conflict have a fifty-fifty chance of finding themselves at war again.[2]

Winning the race against the clock

Patience, persistence and persuasion are undeniably the keys to the success of an intervention; tomorrow's war will be primarily a duel of wills, in which perseverance and tenacity will be cardinal virtues. However, while resolving crises should be seen as a long-term issue, because of the obligation involved, the very duration of the action contains its own pitfalls. The first is that the intervener may find himself trapped by his own intervention. Indeed, even if the use of force can be "justified", it will nevertheless be traumatic for the "intervenee". The longer the engagement, and the greater the scale, the more it is likely to feed reactions of violent rejection. The Coalition has no choice but to engage in a course of reconstruction against the clock while, at the same time, fighting the dynamics of the recurrence of violence. This is made even more difficult by the fact that, once the force has been deployed, it loses a major element of the initiative that had been its initial strategic advantage. It is rapidly forced into a position of reaction rather than action. However, the rule is clear: since it fast becomes counter-productive, the action of the force must be geared to its withdrawal. The only victory being sought will be achieved, at the appropriate time, through the departure of the force.

There is a double paradoxical correlation between the duration of the intervention and its success: staying for a long time does not guarantee success, however leaving rapidly is sure to lead to failure. Struggling among the populations of different cultures, it is difficult for the seed of force to flourish, as its opponents will make every effort to exploit its weaknesses

---

1. According to a report issued by the Rand Corporation (Washington D.C.), between April 2003 and October 2006, Iraq received (after adjustment for inflation) financial aid equal to the sum given to Europe under the Marshall plan – and with what results!
2. Quoted by Naigalé Bagayoko and Anne Kovacs in *La gestion interministerielle des sorties de conflits* [Inter-ministerial management of exiting conflicts], *Les Documents de C2SD*, no. 87, 2007, p. 28.

and faults in the hope of provoking a rejection. Since any intervention is primarily a violent intrusion into the heart of another society, inserting a foreign body will naturally lead it to produce its own antibodies; their growth will only be slowed down by the emergence of tangible proof of the fact that the action of the force is in everyone's interest. In any case, it will be necessary to reduce the length of the stay and its visible marks, to avoid sinking into a decline, followed by the rejection of the force. It is important to delay the point where the army of liberation becomes an army of occupation and then of oppression, and to avoid the proliferation of acts of terrorism transforming the opponent's violence into an insurgency, requiring the permanent presence of a real army. As Gérard Chaliand says: "At a time of nationalism, it requires very little to move from the provisional status of liberator to that of an occupying foreigner."[1]

The Chief of Staff of the French Armed Forces summarised the dilemma confronting the intervention: "While getting bogged down is always a risk, time is an essential ingredient in resolving this type of crisis."[2] His thoughts were echoed by General Petraeus, Commander-in-Chief of the Coalition Forces, when questioned in July 2007 about the conditions for success in Iraq: "None of us, Iraqi or American, are anything but impatient and frustrated at where we are. But there are no shortcuts. Success in an endeavour like this is the result of steady, unremitting pressure over the long haul. It's a test of wills, demanding patience, determination and stamina from all involved"[3]

Projecting forces or projecting strength?

At the core of this problem of time is the question of the mutual benefits of what we in France call the "projection of strength" and the "projection of forces". The first refers to the projection of a threat to destroy or the capacity for remote destruction –with no footprint on the ground, so by air or from the sea– and the second to putting a land force on the ground. In comparison with the situation in the past, when destruction, and thus the projection of strength, were the central capacities of our force structures, what has changed in tomorrow's war is that the long-term aspects and the monitoring of conditions have become prerequisites for the political effectiveness essential to legitimise technical effectiveness. This therefore amounts to saying that –other than in cases where the pro-

---

1. Chaliand, p. 38.
2. Jean-Louis Georgelin, in *L'armée française face aux nouvelles menaces* [The French Army and the New Threats, *Politique Internationale* [International policy], no. 116, Summer 2007, p. 151.
3. In *Valeurs actuelles* [Current values], no. 3688, 3 – 8 August 2007.

jection of strength has a deterrent effect or where, far more unusually, it has a direct political effect[1]– the projection of forces is essential, in order to go beyond technical effectiveness and to achieve political effectiveness. The very projection of strength itself encounters severe problems in deterring and destroying the opponent who, by circumvention, has become asymmetric. While the projection of forces relies for support on the projection of strength, this has become generally incapable of deterring, preventing and stabilising or, in other words, of ensuring by itself the essential strategic "protection" function. Concentrating too much on the projection of strength would also mean condemning ourselves to make growing use of it as, for financial reasons, it would prevent us from having the means to deal with the causes; in other words, we would be condemning ourselves, in the short term, to preventive action.

OPERATING IN THE SAME SPACE AS THE OPPONENT

In the face of the conventional super-strength of his opponent, the adversary's only option is to reject combat using similar weapons and to seek a confrontation which will enable him to achieve political results through very low-level military successes. Applying the eternal rule, he can circumvent the spaces (land, air, sea and electromagnetic) which we dominate, in favour of those where he possesses a comparative advantage by targeting the general feelings of public opinion. Thus he will use the infosphere, which organises the physical separation of strategic and tactical spaces: since tactical operations now take place in a theatre that has strategic repercussions in locations which are geographically very far apart, the opponent is able to project himself away from the physical battlefield, to transpose minor tactical successes to the strategic level and to use these to construct political victories. Thus the battle between the weak and the strong now takes place in the infosphere; the battle in which the opponent operates in psychological fields.

------

1. When this capacity is under discussion, quite rightly the symbolic example always given, because of its uniqueness, is that of Operation El Dorado canyon, which took place on 14 April 1986, which involved a joint raid by the US Air Force and the US Navy on a number of targets in Libya, and in particular in Tripoli. While this rapid surprise raid failed in its attempt to remove the Libyan dictator, it nevertheless completely changed his attitude to the terrorist networks he was using to indulge in a trial of strength with the United States and Western democracies. One might also mention here Operation Opera, which took place on 7 June 1981, and in which some fifteen Israeli fighter-bombers destroyed the Osirak nuclear reactor south-east of Baghdad, as well as another similar operation in Syria on 6 September 2007, the target of which is unknown at the time of writing.

Since it focuses on the sensitivity and versatility of public opinion, communication has become the terrain of victory or defeat. Through communication, where the audience become judges, action presents its results and its acceptability by means of the image it projects. We have no option but to face the opponent in this same space. Inevitably, communication becomes an essential manoeuvre.

Today, winning means being aware that strength is closely monitored by public opinion, and that victory is more likely to be achieved through the acceptance of combat in spaces that "belong" to the opponent than through the detailed planning and tools of the past.

ANTI-TERRORIST OPERATIONS

The changeable and omnipresent threat, no longer bound to a territorial context, and the globalisation of conduct and disputes have undermined the bastions erected by advanced societies for their protection. In today's emotional and irrational world, for the weak, terrorism has become the best way of applying the principle of economy of forces; for limited means and limited risk, it is possible to cause considerable damage, with repercussions across the whole planet. The continuous nature of the threat demands the same of the response and thus increases the value of force structures capable of continuous action where required, using virtually identical resources and methods on either side of the border. The continuum of the threat, in space and in intensity, must be matched by a continuum in operating capacity.

The fight against international terrorism must be seen as a long-term undertaking, since it is likely that this plague will thrive on the disparities and imbalances born of the differences in access to wealth and knowledge for many years to come. However, destroying the terrorists and, possibly also, their sanctuaries –as indispensable as it is– is only one aspect of success. The network structure typical of terrorist organisations makes it impossible to carry out any decisive military operation. Unfortunately, the strategic perception that Western nations have of themselves tends to delay their adaptation to this form of combat which, by its very nature, negates their comparative operational advantages, while centuries of humanitarianism render them incapable of using a whole range of possible responses. The conditions of this fight therefore mean that changes will need to be made to force structures.

Of course, this fight is necessary, but it must be an intelligent fight to deal with an intelligent enemy. Since terrorism –the exploitation of fear through violence in the pursuit of a political result– is more of a mode of communication than a mode of combat. We must place ourselves in the same space, while avoiding giving out any messages –for example, of destruction– which might consolidate the terrorist's action and enhance his position. The term "war on terror" is by no means immune from causing negative effects. Besides the fact that it is difficult to fight a war against a mode of action, this term in itself bestows a degree of legitimacy on the combatants. During the operations in Algeria, from the outset the propaganda referred to members of the FLN as "combatants", thereby giving them, in many people's eyes, a status and a dignity equal to that of the French soldiers. This terminology later wrongly encouraged military action and, implying the notion of "unconditional surrender", removed the idea of compromise, when this would indeed have been possible. It is by no means certain that the words of President George W. Bush on 7 December 2001, the day that marks the commemoration of the "infamy"[1] of Pearl Harbor, implied any sense of strategic direction: "Like all fascists, the terrorists must be defeated. This struggle will not end in victory, not in a truce or a treaty." Furthermore, the expression "war on terror" tends to globalise the fight, while this very globalisation increases the threat and decreases the effectiveness of responses; on the contrary, it is important to adapt the combat to the specific characteristics of each source of contagion.

The fight against terrorism cannot be a war in the conventional sense since destruction cannot be the prime motive. The key can be found in the combination of direct action and sustained, deep-seated action with the need to focus on the causes and underlying emotions rather than on the symptoms. Here, the armed forces can be active in supporting the fight against terrorists or even in counter-terrorism itself –intended to prevent the action and to tackle the very causes of the threat– which is preferable to anti-terrorism, which responds to the attack. Prevention is better than cure, and the former benefits the operations of the troops on the ground. Whereas at home, judicial aspects are more important than military aspects, the dual nature of forces on the ground allows them, outside national boundaries, to combine both types of action to achieve continuity of operations.

It should also be noted that terrorism draws strength from the publicity it receives. It is therefore important to find the correct balance between the measures necessary for the fight and treating such events as commonplace, which has the beneficial effect of reducing the status of the act. If, courageously, societies refuse to allow themselves to be disrupted and are able to find a way to live with the shadow of this curse, it will be even easier to demonstrate its vanity and thus to eradicate it more quickly.

---

1. The expression used by President Franklin D. Roosevelt.

CHAPTER II

AVOIDING MIRAGES

As we have seen, Western nations are finding it difficult to preserve one of their considerable comparative advantages, the military capability to achieve political effectiveness relatively easily. Clinging to the notions of the past, there is a great temptation to seek new weapon systems capable of restoring this advantage, which is being eroded by successive military engagements. However, such a capacity-based approach runs the risk of allowing defence policy to be structured by technology, or even of thinking that war has been transformed by the transformation of weapons, whereas this transformation only alters the terms of combat.

For European nations, there is also a great temptation to follow slavishly the paths of their great American ally and to confuse this model of war with war itself. However, these paths relate to the defence of national interests that are sometimes very far removed from our own, to significantly higher budgetary resources and to essentially technical concepts, which are frequently out of step with the current nature of conflict. Furthermore, these foreign paths are naturally based on a strategic culture and specific patterns of war that also do not fit in with our traditional relationship with war.

It would be dangerous for Europe –and for France in particular– to adopt foreign models with no regard for the culture or the political structure on which they are based, or for their relevance to either European interests, or the characteristics of the likely opponent on the other hand. This is certainly the case for the concepts of RMA (Revolution in Military Affairs) –now known as Transformation– and EBO (Effects Based Operations).

FROM RMA TO TRANSFORMATION

In the United States, Transformation has become a buzzword, referring to the "militarily correct". However, closer examination reveals a contradiction between the apparently brilliant concept and real operational experience.

A vision of the American dream

The word Transformation took on this new meaning in 1999, when General Shinseki, Chief of Staff of the US Army, adopted it in an attempt to rehabilitate the army after the critical analysis which followed the –air only– Operation Allied Force in Kosovo. Shortly afterwards, the idea and the term were picked up by the neo-conservatives, and in particular the team of former Defense Secretary Rumsfeld, as a possible successor to the ageing concept of RMA, which had appeared in American strategic vocabulary in the early 1990s.

RMA is primarily a position setting out a major technico-military change, but it also includes particularly strong political implications. It takes the form of an operation to legitimise the structure and the evolution of the US armed forces: thus, a variety of changes is dressed up and given a "catch-all" name, then promoted strongly at home and abroad. The notion proved to be capable of motivating the military institution, and at the same time, effective at an internal political level in the face of possible decisions to reduce budgets. In terms of the outside world, it acted as an attention-grabbing message, allowing the United States to spread its model around the globe, with anyone refusing to accept this change automatically being classed as an outdated reactionary.

Following the famous speech at the Citadel in September 1999, during the first Bush presidential election campaign, the notion of Transformation became a political weapon to be used against the Democrats, with a view to "ending the military decline". It was then taken up by the new administration, in an attempt to make a break with the previous administration, basing the path of American strength on the ideas of the director of the Office of Net Assessment, Andrew Marshall, and his network-centric designs for military innovation. Contemptuous of the fundamental dialectic aspect of war, the idea put forward by candidate Bush was simple: "The best way to keep the peace is to redefine war on our terms."[1] His vision, which was a technical vision, was of domination through destruction. "Our military must be able to identify targets by a variety of means, then be able to destroy those targets almost instantly, with an array of

---

1. Speech by George W. Bush at the Citadel, 23 September 1999. He later reiterated this idea on a number of occasions, in particular at the naming of the aircraft-carrier Ronald Reagan on 4 March 2001 and at the Naval Academy on 25 May 2001.

weapons, from a submarine-launched cruise missile, to mobile long-range artillery."[1]

Thus, gradually, was born the image of a new invulnerable super-strength, which could be easily deployed in an empty unipolar space, since it was free of all the constraints imposed by multipolarity. To a certain extent, for a single super-strength of this type, the sense of technological superiority enhances political and military freedom of action. The political mechanics of intervention[2] are established at a later stage: in all probability, future operations will be dealt with rapidly, especially if they are carried out in a pre-emptive mode. It is enough to use technology to circumvent the very nature of war and to impose on the opponent a mode of conduct which favours the effectiveness of the available weapons. Alas, it was a mistake to believe that destruction was the same as control, and that the defeat of a state was the same as the submission of a nation. The wars which followed were not information wars, but rather wars fought among the population. Schnaubelt observes that: "The minimal size of ground forces deployed and available for Operation Iraqi Freedom was the result of planning to fight the war we envisioned, with RMA-capabilities we hoped for, instead of the enemy and conditions we would actually face."[3]

The ideas behind Transformation were distributed widely in 1996, in the reference paper Joint Vision 2010, at the beginning of the capacity-based dynamic process, in which the original error was to be a-strategic, in other words, to be detached from any scenario or threat, and thus from any reality. Its authors admitted that they had no idea of the nature of the future enemies of the United States. However, they gave a detailed description of how they would be attacked and how the US should prepare for this combat, which would inevitably be the head-on clash of RMA. They had one simple strategy: "Long-range precision capability, combined with a wide range of delivery systems, is emerging as a key factor in future warfare. [...] The combination of these technology trends will provide an order of magnitude improvement in lethality."[4] It will thus be possible "in the 21st century to find, fix or track and target anything that moves on the

---

1. Ibid.
2. Political sociology will observe that the Iraq War can be seen as one effect of the histori-cal meeting of a current of political thinking and the emergence of new technical capac-ity, offered by RMA. The conceptual framework is that of enlargement (expanding the liberal democratic model) and of shaping (forming the world to render it receptive to American ideas and interests), which over the course of the Clinton era, replaced the former central paradigm of the Cold War, which was containment.
3. Christopher M. Schnaubelt, Whither the RMA, Parameters, Autumn 2007, p. 95.
4. Chairman of the Joint Chief of Staff, Joint Vision 2010, available at http://www.dtic.mil/jv2010/jv2010.pdf, p. 2.

surface of the earth. [...] This emerging reality will change the conduct of warfare".[1] As a result of constant repetition, but without ever having been put to any real test, the maxim of the infallibility of RMA and domination of every opponent solely through the power of Information Dominance gradually became the truth that it was unwise to question. This axiom was particularly well received, as it corresponded to the fantasy of American strategic culture, making it possible, in theory, to conduct a war without getting involved, in other words to escape the dangers of the world while still protecting the American island. From January 2001, network-centric warfare became the sole basis for reflection, operational planning and the design of force structures.

The remarkable ineffectiveness of stand-off operations

Combining, at a safe distance, the progress achieved in precision and information with a structure based in the culture of attrition, the purpose of RMA is really to reconcile the need for war and the aversion to it felt by the American people. The preference given to distant precision bombing, and the predominance of the "fire and forget" style of combat, indicate clearly the relationship with the American strategic culture. It confirms the preferred model of an America dreaming of the sort of invulnerability that an island enjoys and equipped with the means to strike at the heart of any adversary far across the ocean, with total impunity and with no danger of becoming entangled. Operating in a loop from the strategic to the tactical, the RMA model is based on two fundamental ideas: the ability to fire from a safe distance and, in a return to the medieval model of the feudal moat, an impregnable fortress from which a force can be projected to act as the world's policeman.[2]

RMA has undergone a semantic change, to become Transformation – another nebulous term, rather like globalisation, and sufficiently vague to create an idea that can be understood in many ways. It focuses on the question of resources and is made up of the same elements: the considerable progress achieved in the fields of precision, detection and communication, combined with an unshakeable faith that acquiring knowledge will ensure

1. General Ronald R. Fogleman, Global Engagement: A Vision of the 21st century Air Force, available at http://www.au.af.mil/au/awc/awcgate/global/nuvis.htm
2. As Baghdad was taken, Dr Stephen Biddle –of the US Army Strategic Studies Institute (SSI)– described euphorically what he referred to as the "Afghan model": "Bombs guided by special forces commandos, killing at a safe distance [...] all that the local allies have to do, is to protect these commandos from any hostile survivors and to occupy the abandoned terrain at a later stage. The United States can thus defeat these pariah states from a considerable distance, with very few American casualties and very little risk of being seen as a conquering power." March/April 2003, p. 31.

superiority. The performance of the vectors –air, sea and land– is over-shadowed by that of weapon systems interconnected to allow generalised use of the weapon now seen as the most powerful: information. In this vision, Network-Centric Warfare was to enable the American armed forces to move from the industrial age to the "information age". The idea is gradually becoming more firmly developed. While less than 10% of the ammunition used during the Gulf War was precision-guided, this figure rose to 35% in Kosovo and 60% in Afghanistan, with a further leap in Iraq in 2003: 70% of the ammunition consists of smart munitions. Based on the "targeting culture", progress on detection is intended to produce a general, continuously improving overview of the targets to be attacked. For one of the fathers of Transformation, Admiral Owens, Vice Chairman of the Joint Chiefs of Staff, the vision is simple: "What can be seen, can be hit. What can be hit, can be killed."

This approach ultimately leads to the enemy being seen as a collection of targets, while being based on the assumption that these targets exist. Thus, Transformation can only be effective against a conventional enemy, consisting of a collection of material targets which can be detected and destroyed. Nowadays, this is only one of many possible scenarios. It also assumes that the enemy will remain courageously at his post, waiting for the paralysing blows to fall. This is not always the case. In March 2003, the Iraqis, like the rest of the world avid followers of the international defence media, had long since left their installations and dispersed their assets by the time the precision strikes hit the abandoned computers, activated as decoys at the start of the Shock and Awe operation.

Just as the dissymmetric opponent will thus probably be able to avoid the paralysing attack, the same is almost certainly true of our most likely irregular opponent. James S. Corum and Wray R. Johnson observe that: "Generally speaking, guerrillas and terrorists rarely present lucrative targets for aerial attack, and even more rarely is there ever a chance for airpower to be employed in a strategic bombing campaign or even in attack operations on any large scale."[1]

Weapons for war on a grand scale

This vision of a Transformed war reflects American military thinking, which envisages only reluctantly an engagement against any enemy that would not be more or less symmetrical. For the United States, the only war that counts is a "Nations War"; anything else is an "Operation Other Than

---

1. In Airpower in Small Wars, Parameters, US Army, Spring 2007.

War" and can be transferred to subsidiary nations in order to conserve the real strength for engagements in which it can deploy all its technological advantages. The "military-industrial complex" (as it was described by General Eisenhower), which in the United States has a very direct influence on general strategy and operational orientation, obviously has enormous financial interests in this direction.

This typically American approach focuses on destroying the enemy, rather than seeking the best way to achieve a political result. It involves conducting technically perfect attacks, rather than knowing how to transform these operations into a strategic success. This obsession with the battle means concentrating more on its technical and tactical components than on its ultimate aim. Further down the line, defence planning is rooted in the study of capacities, focussing primarily on the equipment needed to optimise mobility, strike power and communications. With no political depth, this capacity-based approach is designed to win battles, not wars.

Thus, in summary, Transformation represents the traditional trans-Atlantic split between political action and military action.[1] While RMA can appear to be a tool to be used "to pursue political objectives by other means", these objectives are essentially internal. For the executive, thanks to the ability to impose its will immediately or –even better– to exercise control from a distance, it is a matter of maintaining the internal political legitimacy of the "war tool", by demonstrating to public opinion that military strength can be used to support American interests, at limited risk and cost and without danger of entanglement. The two major concepts still

---

1. Of course the Americans have read Clausewitz; it is even the standard reference work in American military academies. But they read it as "island dwellers" and re-interpret the thoughts of a man who was essentially a continental dweller of the 18th and 19th centuries, surrounded not by oceans, but by neighbours of varying levels of hostility, and who knew that it was essential to remain in permanent dialogue –armed or unarmed– with them, in an endless game of changing alliances. This is one reason why the American strategic culture introduces the concept of an "interruption" between the political and the military. War is a continuation "by interruption" of politics; it is not really an element of dialogue. In contrast, for Clausewitz, the political aspect is all-encompassing, simply fading for a brief moment, retaining its rights and then returning to exercise them. In his thinking, there is no such interruption, but a continuity from politics to war and then back from war to politics and so on, in an uninterrupted and endless dialogue. Reflecting and underpinning American strategic culture, in the United States, this biased interpretation of Clausewitz shapes the conduct of both politics and war (whereby operational commanders have a high level of autonomy) as well as of the tools of war. To give a recent example, in his excellent book Fiasco (pp. 109, 116), Thomas Ricks refers to what he calls a "strategic disconnect" between the military aim of Operation Iraqi Freedom and its political objective. This topic is developed at greater length in L'Amérique en armes, Anatomie d'une puissance militaire [America in arms, anatomy of a military power] (Desportes).

advocated today (Sea Basing and Rapid Decisive Operations (RDO)) have precisely this aim.

The technological advances initiated by RMA, and implemented by the Transformation process, enable American strategists to envisage the realisation of their ultimate dream: intercontinental precision action, spared from the complexities and uncertainties of ground warfare, conducted from a national safe haven whose land, sea, air and space borders have finally been secured. It is hard for this dream to exist alongside European strategic cultures based on national battlefields, or on those of their neighbours who, in the constantly changing alliances, rapidly became new allies. As Dominique David[1] puts it, in contrast to European countries, "the United States becomes part of the world when it projects itself there, on a non-continuous basis, according to a scheme of comings and goings that favours the criterion of directed effectiveness, rather than the organisation and management of space. [...] With regard to military strategies, we know how this culture of projection and discontinuity organises tactical concepts, the structure of forces and the equipment itself. [...] The strategic movements are organised according to a discontinuous process of establishing a guarantee and force being used in specific locations: a predatory mode."

A theoretical dream, moving ever further from reality

The American difficulties in Iraq and elsewhere have revealed not only the problems of a solely technology and air approach, but also, and more deeply, the significance of prolonged land stabilisation or counter-insurgency operations. At the same time, they make the whole strategic debate in the 1990s, with its focus on decisive victory and increasing pace, seem strangely out of step. The impasse created by politics being forced to take a back seat to technology is suddenly revealed.

From a technical point of view, the problems resulted first of all from the fact that the new opponent was becoming increasingly less visible and thus detectable. This very lack of visibility called into question one of the important pillars of the concept. The idea of the "transparency of the battlefield" no longer applied to the likely opponent. In his excellent book, The Sling and the Stone, Thomas Hammes observed that: "Generally speaking, to the surprise of the supporters of Joint Vision 2020, irregulars have proved that they are now very largely insensitive to our technology."[2]

---

1. Dominique David, *Puissance dominante, puissance référente, hyperpuissance?* [Dominant power, reference power, hyper-power?], speech given on 4 June 2002 at l'Université du Québec, Montreal.
2. Hammes, xii.

The context of current conflicts, comprising force projection and unexpected interventions, as well as prolonged stabilisation efforts and local deterioration, has naturally led to questions being raised by those actually fighting the war, those involved in contact with the opponent, and with violence and death. Transformation seems to function essentially at the level of virtual large-scale confrontations, whereas most real conflicts are mostly at the other end of the spectrum. It will therefore run into difficulties when faced with the reality of tomorrow's wars, the more so since analyses of the two Iraq Wars (January/February 1991 and March/April 2003) and the war in Kosovo (Spring 1999) confirm that, from a technical viewpoint, victory was due less to changes in the form of combat, and more to the crushing disproportion of forces involved.

In addition to this discrepancy between the technical dream and the reality of conflict, the political value of Transformation causes problems. The example of the Iraqi Freedom campaign is particularly clear. In the Spring of 2003, the American "21 day" campaign in Iraq was a remarkable demonstration of the technical prowess achieved by the US armed forces, in spite of the complete lack of balance in the ratio of forces – which puts it somewhat into perspective[1], since a decade of sanctions and regular but localised bombings had severely weakened the Iraqi army. Nevertheless, this impressive tactical and professional victory was unable to create the conditions for strategic success. To this day, the political result bears no relation to what was announced. This situation recalls what Hegel referred to as "the impotence of victory", when describing Napoleon and his conquest of Spain. In November 2005, General Mattis, then head of U.S. Marine Corps Forces Central Command, commenting on this problem, said: "Our fascination with RMA and Transformation has been altered once again by history's enduring lessons about the predominant role of the human discussion in warfare. Our infatuation with technology was a reflection of our own mirror-imaging and an unrealistic desire to dictate the conduct of war on our own terms." He was here challenging the American propensity, often with positive effects, to always try to use a technical solution to overcome a problem. In the face of difficulties arising from then new conditions of intervention, they put their trust in a positive technical asymmetry, intended to resolve the real human asymmetry.

---

1. In *La geurre de demain* [Tomorrow's War], p. 43, Bernard Schnetzler points out that: "The two Gulf Wars, in 1991 and 2003, were an apology for precision weapons. In both cases, critics have neglected to mention that if we calculate the total value of materiel used, from ground level up to an altitude of several thousand kilometres, the opponents were fighting with a force ratio of between ten and a hundred to one. Under such conditions, irrespective of the choice of doctrine and equipment, the question was determined in advance."

The myth of Transformation has given rise to the idea that the tactical benefits that could be derived from RMA were such as to render unnecessary the indispensable link that must be established between the technical results and the political effect. Thus, looking today at the difficulties faced by the superb American war machine, it is essential to examine the basic principles behind, and the ability of, a tool which, while indisputably able to function correctly, has immense difficulties achieving its objectives. The question is not without import, for the training of European armies enables them eventually to fall in with the American model. But we also need to know how to think differently.

A *concept suffering from a lack of political substance*

The example of Iraq is particularly fascinating since it reflects the American tendency to neglect the fundamentally political nature of war. It symbolises the ambivalence of Transformation with, on the one hand, its remarkable success in increasing tactical effectiveness and, on the other, its ambiguity –to say the least– at a strategic level. It leads one to question the point of a tactical victory –and also the technological capacities– that are unable to generate the conditions for strategic success, and to wonder about the concept of a force structure that gives no thought to "the day after". This in turn calls into question the relevance of a philosophy aimed at continuously trying to increase the pace of technical action, when the most likely enemy intends to conduct a prolonged, asymmetric duel and to play the "manoeuvre through weariness"[1] against Western impatience, exchanging losses for time. By concentrating on Transformation –in which technology and its opportunities become an obsession and the means becomes the end– the risk for the advocates of Transformation is that they may find it difficult to think in terms of general strategy, or may find themselves in a position of winning battles but losing campaigns.

We can thus see that Transformation, the result of American positivism, far from guaranteeing the conditions required for strategic success, leads to an impasse. It does not really deliver what was expected; it has not demonstrated any ability to produce the desired political results on the terrain – indeed, far from it. The major crisis affecting the concept which underlies modern Western warfare should automatically lead to questioning not only the conditions of Transformation, but also its general philosophy. The results have failed to materialise. Not in technical terms –indeed, Transformation does seem to have a greater destruction capability– but in terms of strategic results. Today's opponent has understood that it is in his interest to

---

1. Beaufre.

present our high-performance weapons with either an empty battlefield or one that is very much interwoven with the local population. The "transparency" of the battlefield only relates to objects which can be digitized. In an asymmetric war, these are neither the majority nor essential. In July 2006, the Israeli Air Force very rapidly ran out of "military" targets that could be profitably destroyed. Former Defense Secretary Rumsfeld publicly lamented this same situation in October 2001, just a few days into the air war in Afghanistan. It is naturally in the opponent's interests to reject the type of combat the US Air Force would like to impose on him.

The gravest error we could commit would be to concentrate our defence efforts too much on a single type of system, for example, on stand-off precision strikes. The uniformity of the threat would remove all uncertainty for the opponent, who would have no difficulty in exploiting this and devising a form of defence and attack that would catch us unprepared. It would be far better for Western countries to avoid being drawn into some form of Maginot Line.

RMA, victim of the eternal rule of circumvention

Historically, each Revolution in Military Affairs sooner or later provokes avoidance counter-measures, and Transformation is no exception to this. There is also a dual phenomenon: firstly, the enemy less frequently appears as a target (having understood that if he presents a target, he will be destroyed) and secondly, destruction is proving counter-productive in the search for direct political effectiveness. Iraq has demonstrated clearly that, despite having been a remarkable tactical victory, in which the OODA loop has never been shorter, the three weeks of war in March and April 2003 produced no positive strategic result at that stage. There was a technical result, but this ultimately went no further.

Right on cue, the Israeli actions in the summer of 2006 confirmed this verdict. Despite the irrefutable technical results, in terms of the destruction of targets on the ground from the air, the IDF were incapable of achieving their initial political objective. Technical perfection alone was insufficient to produce the desired end state. Are we looking for a swift result and a minimum engagement? On the contrary, we are moving towards prolonged land operations. Iraq has been three and a half weeks of "Transformed" war and well over four years of asymmetric crisis. After two and half years, American strength had dealt with Germany, Italy and Japan. How far have we got today in Iraq? And Kosovo? 78 days of "Transformed" war, or war in the course of "Transformation", for no clear political result. We have now entered the ninth year of "non-Transformed" crisis. "Transformed" war does not seem to make it possible to pursue a political aim effectively by other means. It stumbles on the weaknesses of its initial concept (focus on technology and combat, confused with war) and the certainty that this represents primarily the destruction of the other side's combat strength and the neglect of political factors.

Following a decade of evolution intended to perfect what was already a high-performance tool in preparation for an increasingly unlikely engagement, Transformation can now be seen for what it really is: a project for a theoretical war, in most cases bearing very little resemblance to reality. It follows its own internal dynamics, existing solely for its own sake, removed from the constraints that would be imposed by any likely enemy, particularly since it is only technically feasible because it is possible to derive some significant political or strategic benefit. It is perhaps somewhat surprising to note that, while we expect future wars to be prolonged affairs, we continue to invest billions of euros in attempts to shave a few seconds off the decisional loop. It is also astounding that, on both sides of the Atlantic, the major "Transformed" war exercises are conducted against an opponent who both disregards the fundamental rule of circumvention, and is "not Transformed", while the very nature of Revolutions in Military Affairs is that they end in a spreading of their effects increasingly rapidly.

Observing the current chaos in Iraq, which seriously discredits the concept of Transformation, it is clear that Operation Iraqi Freedom was explicitly devised as a means of validating the concept of Transformation and was immediately seen as immediate proof of its validity. From the very first moments of the campaign, Defense Secretary Rumsfeld put all his weight behind it, determining the conditions and details in order to validate the concept of Transformation by demonstrating that it is possible to substitute speed for mass, and that Information Dominance allowed US forces to dispense with the rules of a bygone age (legacy thinking). For his part, following the fall of Baghdad, Vice President Cheney declared this victory to be "proof positive of the success of our efforts to transform the military". Speaking at a hearing of the Senate Armed Forces Committee, Secretary Rumsfeld and General Franks –the commander of the operation– presented it as a definitive validation of the concepts underpinning Transformation.[1] Later, but taking a critical view, the analyst Greg Grant observed that: "In what was supposed to be a new approach to combat in the 21st century, with a revolution in military affairs abolishing close combat, the reality of the battlefield delivered a sizeable dose of counter-revolutionary reality, the lessons of which we are only just starting to learn."[2]

1. Kagan, p. 346.
2. Defense News, 9 October 2006, p. 30.

Transforming Transformation

We should be in no doubt that RMA, and its successor Transformation, are a true reflection of a specific strategic culture, that of the United States.[1] In particular, ever since the Civil War, the conditions under which the American hyper-power has been used have enabled the United States to employ force in what has usually been a very favourable ratio of superiority. This ease has naturally led to a tendency to under-estimate the reactions of the opponent and to devising modes of operations that would be unthinkable in situations of parity. The problem with such power is its tendency to obscure reality. It encourages neither strategy, nor refinement; there is no need to understand the other side if you can easily create a path by means of force, and contact is not technically necessary to eliminate the opponent. For Americans, another negative effect of such power is the size of the bureaucratic machine; it is already so difficult to reach a consensus among all the parties involved internally in any decision, that listening to what the other side is saying becomes almost impossible.

One of the characteristics of American strategic culture is the desire to try to bend the rules of war to take account of technological innovations designed to remove physical constraints. In 1944, R.P. Patterson, Under Secretary for War announced: "We will ensure that our plans suit our weapons, rather than our weapons suiting our plans." This expresses a continual desire to try and circumvent a fundamental rule of war –that of reciprocal action– by saturation (annihilation), speed (e.g. OODA), action that precludes reaction (e.g. Shock and Awe) or actions before the action (pre-emptive action). However, all this assumes that the opponent is vulnerable to the weapons available; as soon as a response is found, the idea is confounded and armies are obliged to return to the inescapable, and timeless, truths of warfare. The current trend is to construct a virtual opponent who directs efforts towards the information war. After investigation, we come to believe in his existence or in the probable return of a "peer competitor", a close genetic relative; in other words, a champion on the other side who would be prepared to join battle in the closed world of high-technology. This is at the cost of the real opponent, who adapts and avoids. Shaped by Transformation, the world is viewed through a technology-based analysis matrix, which defines intervention capability and, ultimately, the ethics of international conduct. In some ways, concentrating on the search for what the Americans call the "silver bullet", either technical or conceptual, able to ensure instant victory in all cases, means that the political dimension of war and the need to develop an internal strategy are liable to be forgotten.

---

1. For more on this subject, see *L'Amérique en armes, Anatomie d'une puissance militaire* [America in arms, anatomy of a military power], Desportes.

The art of warfare is a tangible art. The relevance of a concept or a strategy is only revealed by its results. From the standpoint of today, Transformation appears as a perfectionist response to a question that is no longer being asked.[1] If we wish to wish to continue along the path of Transformation –a route which seemed so attractive in the early 2000s, but which is now increasingly being called into question in the United States as well[2]– it is essential to review it in the light of our own strategic culture and our vision of warfare. We need to examine the gulf between yesterday's expectations, which are still being followed, and the way in which it should be applied today. We must understand what the new technical achievements are not able to provide. We must also re-orient Transformation towards the new reality of conflict and give the same priority to limiting civilian casualties and damage as we do to destroying the enemy.

Transformation focuses on combat, but war is much more than that: it is first and foremost political. Thus, by concentrating too hard on the "forms of war", Transformation neglects the essential aspects, which are by no means simply the techniques of war. Transformation is concerned with fighting a "better war", whereas war in general, and tomorrow's war in particular, are aiming for a "better peace". It is essential to give this technical approach some political substance and to look firstly at the ultimate goal. RMA cannot be solely military; it must be based on not only technical and tactical considerations, but also on global and political concerns, including all aspects of national strength.

It is probably also time to abandon forever the notion of Transformation, which is vague, tarnished and, as things currently stand, synonymous with failure. We need to think differently.[3]

---

1. Expression used by Jean-Michel Millet in *Contre-révolutions dans les affaires militaires* [Counter-Revolutions in Military Affairs], *Inflexions*, no. 4, December 2006.
2. One of several authors who has written on this subject is Etienne de Durand, *Les relations civilio-militaires américaines depuis 2003* [American civilian-military relations since 2003], a consultancy paper for the *DAS* (Strategic Affairs Delegation), December 2006.
3. In passing, it is worth noting that it was only the Prussian general Scharnhorst's belief that officers should be independent, in order to be able to grasp and operate in the new conditions of war imposed by the French revolutionary model, which allowed him to bring about his Copernican revolution (enabling the Prussian armies, which were severely traumatised by the defeat at Jena (1806), to understand the changes in a world in which they had lost their effectiveness, and to achieve a radical transformation in their thinking). He believed that, to acquire the new intellectual and technical tools that would be required, it was essential to go beyond the prevailing thinking. Thus the modern Prussian Army was born out of the intellectual revolution Scharnhorst initiated once he understood that the face of war had undergone a profound change, meaning that the intellectual paradigms which corresponded to the previous forms of war needed to make way for the new military effectiveness. France has had to pay a heavy price for his intelligence.

FOOTBALL OR SOCCER[1]?

Seemingly new operational concepts do not just appear out of thin air. On the contrary, they mark a new, and inherently transitory, stage in the gradual maturing of strategic culture. They thus also reflect the culture and characteristics of a nation. Transformation is not only indicative of American strategic culture, but also of its social culture, a key aspect of which is the role of baseball and American football[2].

It is interesting to note that, since the middle of the 19th century, the American model of combat, with the emphasis on "a crushing force"[3] whatever the form, and even today with Transformation, reflects the spirit of American football, which, however, seems ill-suited to tomorrow's war. One can find the same principles for management and centralised execution, the fundamental idea of assembling forces to break through enemy lines and the notion of a concentration of force within a strict framework of rules. In contrast, soccer teams avoid concentration and players are dispersed across the field to take advantage of open spaces. There is a whole palette of threats, surprise attacks and patient waiting, with individual initiatives constantly reshaping the game to seize the opportunities offered. In soccer, it is not strength that counts, but rather finesse, continuous movement, deceiving the enemy and maintaining uncertainty in order to create strategic chances. The ball comes and goes, rises and falls, waiting for the moment. The principles of Sun Tzu are applied to this fluid game; creating the "potential offered by the situation", waiting for the favourable moment, avoiding the lines and the concentrations of strength. Attacks are hidden, indirect and unexpected, decided on rapidly by the players involved. There is no real central plan or "game plan" determined in advance during training. The game remains open and the defender never knows where the attack will come from. These are the tactics of the irregular adversary who patiently bides his time, deceiving and misleading his opponent, before scoring a goal when the time is right.

---

1. Soccer is what the Americans call the game Europeans know as football. When an American refers to football, he means American football, a game in which the players wear helmets and full protection and pick up the oval ball with their hands, more akin to rugby.
2. The excellent article by Joël F. Cassman and David Lai in Armed Forces journal, November 2003, pp. 49 and foll., sheds greater light on this.
3. This refers to what has been called the "Grant paradigm". A biased reading of Clausewitz and Jomini combines with the founding experience of the Civil War to produce a military culture dominated by the search for the total annihilation of the adversary, in a decisive battle between equal sides, in order to obtain a surrender as quickly as possible.

The American armed forces are very keen on American football. It is typical that the most important sporting contest between the three service academies (West Point, Naval Academy, Air Force Academy) is the American football competition. Similarly, there is a crossover between the two vocabularies: footballers talk of "blitzes", "trenches" and other "bombs", while the Vietnam War had its "Operation Linebacker"[1] and the First Gulf War its "Hail Mary manoeuvre"[2]. The problem today is that, while the spirit of American football has proved effective against traditional forces, it would be difficult to apply it to a loosely-defined, dispersed adversary using soccer tactics. The power of the "shock" is neutralised by the primitive yet effective techniques, as it was in the battle of Mogadishu (1993). It is still being used ineffectively in Iraq and Afghanistan, as well as in other places where the irregular enemy is using his tactics on the world's soccer pitch.

The purpose of this ball game analogy is to try and point out the need to weigh up the pros and cons before adopting elements which come from a strategic culture which differs from ours and whose constituent elements are perhaps not well suited to the wars we will be called upon to fight. We need to operate and we need to operate with them, while nevertheless thinking otherwise.

ONE OPERATIONAL DREAM: EFFECTS BASED OPERATIONS (EBO)

Under cover of first RMA, then Transformation, the natural optimism of American doctrine has devised a series of operational methods shaped by the strategic culture of wealth and strength. Thus the American mentality favours rapid, decisive results and finds it difficult to understand the ambiguous approaches to restricted combat. Moreover, since American democracy will only be prised away from its internal preoccupations in the event of something major happening, if the USA decides to wage war it does so completely, in other words, with all its strength, down to the destruction –at least politically, but sometimes also physically– of its opponent. General MacArthur's statement that "in war, there is no substitute for victory" is the summary of experience and the basis for doctrine: the obsession with victory is a recurring theme, as are its implementation, the concentration of resources and the effort aimed at achieving a decisive and rapid result

---

1. Aerial bombing operations (Linebacker I and Linebacker II) carried out in 1972 to halt logistic movements between North and South.
2. The plan initially put forward by General Schwarzkopf, commander of the Coalition forces, for the liberation of Kuwait.

through a direct offensive. The golden rule, "Achieve victory early" –that benchmark of effectiveness– determines the ways and means, drawing on the best that technology has to offer. The enemy must be destroyed because strength makes it possible to do so and because it is the shortest route to military victory. Of course, this military victory is then expected to be translated into political victory and a return to peace before national public opinion has even had a chance to draw breath. It is this desire which has a fundamental influence on the strategies and forms of war.

The origin

Today, all these operational methods are intended to derive as much benefit as possible from the comparative essential advantage the United States possesses –the most advanced technology and its multiplying effects– and thus to achieve rapid success in wars based on the new model: to overwhelm the opponent immediately by means of a strike intended to "dazzle", "paralyse", "transfix" or "decapitate", etc., and thus to crush the clearly identified troublemaker, at the least cost, particularly in terms of American lives. American military strategy has always been more concerned with achieving a decisive victory than with uncertain political goals. Such methods are interested not in a desire to negotiate or to compromise, or with action linked to the causes of the problem, but rather, in contrast, in overcoming as quickly as possible the disruption of world order, that is, of the market democracy in which American interests prosper. Thus we have seen –each one more virtual than its predecessor, having never been required to confront reality– the successive concepts of Decisive Force, Preclusion, Rapid Dominance, Rapid Decisive Operation and EBO, the first version of which appeared in the mid-1990s. In all these cases, according to the excellent summary given by General Shinseki, the American armed forces wanted "to initiate combat on their own terms, at the time and place and according to the methods of their own choosing."[1]

The case of EBO is particularly interesting. It is, in fact, a typical example of a method invented to defend corporatist interests, then imposed, by virtue of strength on outside observers only too willing to seize any hope of an easy solution to the existential problem of the incapacity of conventional forces. What a relief it is to discover the "silver bullet" in the allied arsenal, especially if decorated in the brightest technological colours! Nevertheless, the question must be asked as to whether this new EBO formula will dispel the uncertainties of Transformation.

---

1. General Shinseki. Chief of Staff of the US Army, Senate hearing, 25 April 2000.

In practical terms, this concept, which started in the US Air Force as Effect Based Targeting, does not advance us a great deal. It was introduced by General David Deptula in an attempt to salvage the USAF by distancing it from its only technique, attrition or, in other words, the progressive exhaustion of the opponent through systematic destruction. Rejecting the concentration on statistic indicators used to measure the effectiveness of air forces (quantity of bombs dropped, number of aircraft available, number of sorties carried out, etc.), Deptula established a theory focussing on the results –in particular in the cognitive field– rather than on the means and the quantitative approach of the "body count". Unfortunately, the rejection of numerical indicators ultimately led to a systematic approach, reducing the practical relevance of the excellent initial idea.

Since war is, by its very nature, an instrument of psychological action, using effects intended to encourage the enemy to behave in the way we would hope is, of course, not a new idea. However, it made a comeback during the 1990s, when the euphoria of the stunning victory of the First Gulf War combined with the extraordinary technical possibilities offered by new precision and communications technology. There was a wish to extend the "targeting" to the whole spectrum of operations. EBO –this catch-all expression, with as many versions as there are major commands using it– seem at first glance to be quite ordinary, but with a very alluring title. They are, in part at least, the rediscovery that the effect, or the end, is more important than the means. In most cases, however, the concept of "effect" is wrongly defined; it is confused with the idea of "aim" in an analysis which regularly ignores the fundamental difference between the purpose of war (the *Zweck*) and the goal of war (the *Ziel*), to refer once again to Clausewitz's excellent distinction.

To understand the Americans' new attraction to an approach that strikes us as commonplace, we should remember that their tactical reasoning did not include the concept of effects. Rather, it was based on a succession of tasks to be carried out. By contrast, in French military thinking, tactical and strategic reasoning has always been based on the concept of effect, organised around the most important effect, the point of convergence in the action.

Dangerous illusions

The concept of EBO comprises a number of dangerous illusions.

The first –somewhat scornful of Clausewitz's theories, but following directly from those of Admiral Owen– is that the conduct of war can be meticulously orchestrated. Having supposedly overcome the "fog of war" and the extreme complexity of the management of effects, advance knowledge of the opponent and the consequences of the action would seemingly make it possible to achieve measured effects, unfolding in a rational and controlled manner. The subsequent cascade of actions/reactions would then make it possible to reach the desired end state by means of a tightly controlled sequence of events, taking account in real time of the results obtained. It is, of course, impossible to reduce the extremely wide range of interactions that constitute combat to a mathematical model. Furthermore, while its rationality would appear to be an ideal solution, the overall control required by the process is so complex that it excludes any flexibility or adaptation to the inevitable changes in the circumstances. Any rapid reaction automatically means diverging from the model, which does not allow for any swift re-evaluation of the hypotheses and observations made in what is essentially a highly volatile environment.

The second illusion is that the conduct of individual, and the way individuals behave as a group, can reasonably be predicted. Sadly, in real life, actors do not always follow the paths predicted by specialists in modelling theory. The reactions of the opposing political decision-makers may be based on considerations which are very foreign to our rationale, thus making them difficult to identify, predict or analyse. According to the EBO concept, if we are aware of the impact of physical effects on moral, we should be able to control the reactions of the other side with no difficulty, something which is clearly refuted by many centuries of military history. Professor Paul K. Davis, research director of the most important American think-tank, the RAND Corporation, is one of many who have expressed this opinion: "My most radical criticism of the approaches favoured by EBO, is that they are based on unrealistic attempts to predict accurately the consequences of a mode of action."[1] Here Davis was simply re-iterating one of Clausewitz's key ideas, borne out on many occasions: "In war more than anywhere else... the disparity between cause and effect may be such that the critic is not justified in considering the effects as inevitable results of known causes."[2]

A further problem is the difficulty of evaluating the results of the effects; continuous evaluation of the operation as a whole must be possible, in order to adjust it where necessary. There is an inconsistency between the very idea of a rapid victory (thanks to EBO) and the time needed to identify the effects. In these applications, the effort of re-plan-

---

1. Interview in *Défense et Sécurité Internationale* [International Defence and Security], No. 27, June 2007, p. 50.
2. Clausewitz, p. 156.

ning and the desire for constant acceleration tends to lead to the suppression of the Observation phase of OODA; the machine gets caught up in the desire for excellence, becoming a machine that strikes increasingly rapidly, but also increasingly at the wrong time, with a counter-productive effect.

A disconnect between physical strength and psychological effect

It is not easy to evaluate the link between an event and its psychological effect, particularly with regard to the time taken to establish the psychological effect. The unalterable time of psychological latency means that it is not possible to envisage a rapid succession of psychological effects in the same way as a rapid sequence of material and tactical actions against a traditional opponent.

Even more fundamentally, in terms of the relationship between physical and psychological effects, the military history of the 20th century shows us that most of the great military theorists were simply wrong: we must now accept that the relationship between military strength and the psychology of the adversary cannot be easily predicted. Furthermore, another idea we have long held to be certain now seems increasingly open to question: once a certain level of superiority has been achieved, however this strength may be deployed, that side will win.

The notion of aerial strength –the concentration of industrial-scale warfare– allowing victory and leading to direct strategic success without crossing swords is an idea which dates more or less from the first flight by an armed aircraft. It has been advanced by Guilio Douhet, Sir Hugh Trenchard, Billy Mitchell, Alexander de Seversky, Warden and others; sadly, since it reveals a lack of understanding of war, it cannot work. Without exception, the initial shock does not initiiate a psychological process that leads to the opponent capitulating. The air campaign Instant Thunder was not enough to convince Saddam Hussein to withdraw from Kuwait; nor does it explain the success of the ground attack on 24 February 1991. It was the combination of the two –in particular the constant threat posed by the coalition land forces, which led the Iraqis to concentrate their armoured troops, thereby leaving them vulnerable to attack– which produced the result. Contrary to the expectations behind the American Shock and Awe theory (devised in 1996 to give American forces the critical Rapid Dominance[1] capability), which was intended to convince the Serbs, in 1999, and the Iraqis, in 2003, to capitulate as soon as the first wave of

---

1. Harlan Ullman and James P. Wade, Shock and Awe: Achieving Rapid Dominance, 1996.

attacks was launched, both countries continued to resist until a certain level of attrition had been reached. This resistance lasted 78 days in 1999 and three weeks in 2003. The war in Kosovo –the "limited operation"[1], which nevertheless mobilised over 700 US aircraft, a third of its military fleet– initially seemed to demonstrate that air power alone could lead to victory. However, later analyses have shown that the lack of synergy with a land threat was the primary reason why the air campaign lasted so long, and that the Serbian defeat was the result of a combination of factors, including the withdrawal of Soviet support from Belgrade, the role of the Kosovo Liberation army and, finally, the threat of a ground invasion. It is not shock that leads a side to give up, but rather the sense of the futility of resistance and the loss of strength... unless, of course, there is someone like General de Gaulle to "grasp the sword"[2]!

Whatever Douhet, Mitchell and the architects of the air weapon may have thought about its ability to achieve a rapid victory by itself, there is no link between the amount of explosives dropped and the breaking of the other party's spirit. The hundreds of thousands of casualties in Guernica, Coventry, Dresden and Tokyo did not alter the desire of their nations or their governments to fight and to resist – quite the reverse, in fact. General Falkenhayn's guns in Verdun were unable to succeed in wearing down the French powers of endurance. Air Force General Marxhall Harris's brutal and indiscriminate bombing of Germany failed to produce the desired effect. Similarly, neither General Le May's 44-month bombing campaign against Hanoi nor General Clark's strikes in Serbia yielded the expected results.

Today historians are in agreement that the actual effects of bombing raids should be examined with great caution. Their effect on morale seems universally to be almost negligible. According to Hervé Coutau-Bégarie, "the psychological effect is obvious"[3], *[is this right? Seems to contradict the previous sentence]* while the official historians of the USAF believe that the effects of strategic bombing were "gradual, cumulative and seldom quantifiable with any degree of confidence throughout the entire campaign".[4] A closer examination of the limited results of the strategic bombing raids during the Second World War, their ambiguous nature during the Korean and Vietnam wars and their inability to bring about the fall of Saddam Hussein, either in the Gulf Wars or in the decade that followed

---

1. Small Scale Contingency!
2. Speech in London, BBC, 13 July 1940.
3. Coutau-Bégarie, p. 619. He is here referring to the strategic bombing of Germany in the Second World War.
4. Murray, p. 112.

the First Gulf War, leads to serious questions about the effectiveness of this strategy, at least with regard to the psychological dimension. If it is directly targeting the civilian population, it does not seem able, by itself, to bring about the desired collapse of morale. It even appears that the time needed to produce the expected cognitive effects –the conflicts in Kosovo (78 days of operation rather than the 5 initially predicted) and Lebanon (33 days rather than 2) demonstrate that this period is always relatively long– works against the democracies, which may lose patience even before their opponents.

Negative side effects

In addition to the inherent difficulties of implementation, the mechanistic concept underlying EBO takes no account of friction. Its intention is to exercise detailed control of the chains of effect, reactions and counterreactions, and it conveys the dangerous illusion that it would be possible, in real time, to have full and exhaustive knowledge of the opponent and the environment. It also assumes a detailed knowledge of the theatre and of the functioning of a complex system, despite the constant changes. One cannot help but think that this concept severely underestimates the opponent, and wonder why, with such excellent knowledge of the theatre, it was not possible to prevent the crisis in the first place!

In terms of command, the effect of this process –generally permeated by the negative side-effects of OODA– is twofold. Firstly, it imposes a centralised direction of operations, at the expense of freedom of action for lower level commanders, which is so crucial in the new, changeable patterns of conflict. Since it favours the "sensor-to-shooter" link, the resulting structure minimises the observation and evaluation aspects.

Secondly, this process, conducted scientifically by the staffs, considerably reduces the role of the commander, with command gradually becoming simply management. As this can easily produce a disconnect from the reality of theatres of operation, the method has a tendency to return to mechanistic concepts of war, and the failures we are all familiar with. It thus suffers the weakness of having been designed for use against a single opponent, organised into a system along hierarchical lines and with a rationality that can easily be influenced. By contrast, the enemy we are most likely to encounter will be multiple and unlikely to be thwarted by decapitating strikes. Furthermore, the method is unrealistic; it assumes that it is possible to coordinate all the actors –military and non-military, national and allied– and thus to control all resources, including those of independent organisations. This results in a probably ideal, yet strangely fragile, approach. Today, at a time where the political absurdity of a tactical demonstration has been clearly established, the fact that American commanders saw the tactical success of the initial campaign in Iraq as a "life-size" validation of the concept raises profound questions.

Rejecting fashion for fashion's sake and knowing how to adapt to the "French way"

Offering little that is new to French tactical thinking (as least as far as land forces are concerned), EBO is one of these new fashions which, after reshaping the outlines and renaming principles that have generally already been tested, believes it can offer an elegant response to our strategic anxieties. They are often cunningly presented and made to look quite brilliant; a gullible observer would be even more easily mystified by giving in to the temptation of ignorance, that of accepting the easy notion that each war is new. The means are new, the ways are new, but the nature of war remains unchanged. The temptation offered by ignorance, caught up in the speed of change, is to think that past experience has nothing of value to offer. However, anyone taking the trouble to read through the work of military authors of the past two millennia will discover two apparently contradictory phenomena. The first is that, for centuries, we have had the feeling that we are fighting new wars, unrelated to previous conflicts. The second is that, in contrast, with the benefit of hindsight, it is surprising to see the stability of the general characteristics of conflicts, their unchanging logic and the error that could have been avoided if the "trendsetters" of the period had simply had longer memories.

The textbook case provided by EBO is particularly interesting. It teaches us much, about both the remarkable ability of the American armed forces to come up with new operational methods, and the cultural foundation for these brilliant technology-centred constructions. It is easy to understand that this bold intellectual construction, which can readily be translated into "instrument panels" or multi-coloured slides, may appear very attractive to those of a fundamentally positive frame of mind. The case of EBO also teaches us that, despite the restraint we must show before committing ourselves hastily as a result of the combined effect of influence and admiration, we should endeavour to adopt the best elements from trans-Atlantic thinking; the US armed forces possess, in numerical terms, a capacity for reflection and, in cultural terms, a capacity for innovation and questioning that we do not have.

This method of establishing the cohesion of the totality of actions in a given theatre comprises clearly positive aspects, which we must make use of in the way that best befits our national character – *à la française*. For many people, it has already brought about a re-acquaintance with the non-military dimensions of war and the need to coordinate outside the purely military world. Having developed the required distance, what the principle of EBO offers us is the ability to go beyond our, sometimes rather strong, Western relationship with a technology-based approach. It is this relationship which has ultimately caused us to forget that the real question relates to the purpose of the military action.

CHAPTER III

A GLOBAL OPERATION

The wars we are likely to be called upon to fight, because we will have been unable to prevent them, will demand integrated politico-military responses: there are, in effect, no longer any simple and decisive military solutions to the world's problems. Success depends on adopting a global approach.

By virtue of their nature, the armed forces will play a role in the most difficult conditions, since they will not become involved until all else has failed, which, to a great extent, explains their recurring difficulties in the various theatres. In these difficult situations, military intervention will have to take account of the pressing need to ensure that the actions of all the parties involved –inter-ministerial, non-governmental, private– at various levels –strategic, operational and tactical– are closely coordinated. The implementation of the simple principles of unity of purpose and, if possible, of overall management becomes a crucial condition for success.

War is many things; however, it is never simply a matter of pointing a weapon at a target. Since the effect being sought is ultimately a political effect, the precedence of the political approach over the purely military approach must be the norm. In tomorrow's war, while military instruments may be able to contain irregular violence, only political means will be able to defeat it.

THE CRUCIAL ROLE OF INTELLIGENCE

In future wars, intelligence will become more vital than ever before. This is particularly true of strategic intelligence, ahead of the crisis and the interventions, because intelligence is the condition which determines the anticipation and the evaluation of the situation, and thus of prevention and intelligent intervention. As a result of the considerable difficulties currently met by forces carrying out their missions in the new contexts of engagement, and the growing rejection of "interference", prevention has become an absolute priority for the strategic function. As things progress, monitoring the general situation, following international developments and the most vulnerable areas, and taking action leading up to the crisis are by far the best and most economical solutions. Intelligence and information –and thus the ability to anticipate– have a key role to play.

Tomorrow's wars will be intelligence operations

Once the intervention has been launched, reality brings us back from the virtual war and the transparent battlespace on our screen to the realities imposed by the terrain. Tomorrow's war will, at best, be conducted like an intelligence operation –it is only partly military in nature– and not like a destruction operation, which is how wars were fought in the past. The first task for the armed forces is an intelligence and information-gathering task, for use not only against the adversary, but also in order to be able to "work with" him. The gathering, processing and dissemination of intelligence should also make it possible to carry out proactive actions, such as identifying the expectations and wishes of the population, in order to be able to convince people at the earliest possible stage how the interference they are being subjected to will benefit them. Arriving in a theatre of operations means becoming part of a situation in which understanding is the key to success. Intelligence is a high priority, covering a range of factors.

The clear picture of a known opponent, concrete targets, strategic visions, military goals and open spaces has been replaced by the uncertainties of the constantly changing "chameleon" adversary[1], the human environment, combat in intangible fields, micro-actions at a low tactical level and confined spaces. In the past, intelligence was directed more at targeting than at decision-making; in view of the new circumstances a re-orientation is required. It is no longer a matter of organising a successful common action involving a limited number of units acting against a clearly identified adversary with a clear purpose, but rather of ensuring that the amalgamation of the actions of a large number of small teams contributes to the success of the whole, despite the fact that these teams are highly decentralised and are facing a little known opponent across a dispersed zone of action. The "contested zones" in which today's land engagements take place are characterised by a high degree of heterogeneity, both physical and human. In these areas, information superiority is greatly reduced, communications systems are stretched and the rugged environment renders observation and interpretation very difficult.

---

1. As Clausewitz put it so well.

Useful intelligence is a subjective notion

To a great extent, the nature of intelligence seems to have changed from being objective to being subjective. Yesterday, the emphasis was on gathering concrete information; the processed intelligence could be readily presented in the form of tables, organisational diagrams and graphs. Nowadays, the true goals of intelligence are intentions, which cannot easily be deduced from situations. Understanding the complex dynamics of the threat and the overall picture requires just as much insight and intuition as it does rationality. There is a shift from the deductive to the intuitive; rather than progressing from the facts to the idea, we must move from the idea to the facts in order to discover them, and thus from an understanding of the context to defining useful targets. We need to understand in order to know and to analyse the opponent's purpose, and thus his possible modes of action, rather than simply using standard methods as a matter of course. On tomorrow's battlefields, the concept of "Information Dominance" will have lost much of its sense, as it will no longer automatically guarantee superiority. This is demonstrated on a daily basis by the difficulties encountered in Iraq and Afghanistan, where many of the concepts and systems of industrial warfare have proved worthless. This situation is reminiscent of a game of chess: the two players have exactly the same "information" about the situation, but the winner is the one with the greater understanding of the situation and of the psychology of his opponent. In other words, what will count in future wars will not be "Information Dominance", but "Knowledge Dominance" and, even more so, "Understanding Dominance".

Under these conditions, it is clear that intelligence has become both more important and more complex, and that even the most up-to-date of equipment and systems designed for past situations are often either inadequate or indeed completely unsuitable. We are truly facing a revolution in this area. Failure to act on this would mean condemning our forces in the future to fight blindly and reactively against an opponent who holds the initiative and, consequently, all the information needed for his actions. It is quite evident that changes in military action require a corresponding change in intelligence. General Fast, commander of the US Army Intelligence School, stated this clearly in December 2006: "Operations in Iraq and Afghanistan have made necessary a readjustment from a 'sensor-centric' approach to a system whose prime purpose is to provide intelligence support adapted to those who have the greatest need of information: the soldier and the small unit commander."[1]

---

1. Quoted in *Doctrine* no. 9-2006, "*Renseigner pour les forces*" [Intelligence for the forces], Défense, CDEF, Paris.

Priority must be given to small tactical units

Overall, the problem is a complicated one: for the most part, what was needed in the past is still needed, as it is necessary to prepare for a traditional war, if we want to prevent it happening, and also to prepare for its resurgence, but in addition the probable future wars also impose new demands. The new opponent is increasingly difficult to detect, and thus to locate and to attack using precision weapons. In the past, the core of military action was destruction, preceded by intelligence, while nowadays, understanding and knowledge of the situation are essential, with the identification of micro-situations and micro-objectives. The "transparent battlefield" increasingly seems to be a questionable theoretical notion. It was thought to be possible to fight "using" information; now we understand that we also have to fight "for" information.

There is a shift in priorities across the various levels: the strategic level is making way for the tactical level and there is a move from a "top down" approach to a "bottom up" approach, where the troops in contact[1] provide the mass of uncertain indications which serve as a basis for the operational analysis. Consequently, the process and the arrangements are reversed. The lower tactical levels need more means of acquisition and a greater capacity for analysis. We need to take another look at digitization which, in this area, was designed for yesterday's traditional war, with its "top down" approach, and which is now not as useful as when we were expecting to be attacking concrete targets.

Once again the "fog of war" has thickened across the battlefield. The US Army, confronted directly in Iraq by the new reality of conflict, is now becoming aware of the clear risk of knowing more and more about one's friends and proportionally less and less about one's enemies. It understands the danger, at the now dominant tactical level, of the impossibility of merging electronically the "blue" images –always accurate and exact – with the "red" images– always inexact and late. It is beginning to call into question the Common Operational Picture (COP), which is different from the Real Operational Picture (ROP).

---

1. Two of the ten directives issued to his units in June 2007 by General Petraeus, commander of the Coalition forces in Iraq reflect this development. One reads: "We are in a fight for intelligence – all the time." In the new American vision, there should be no waiting for intelligence to come from higher levels, but rather all tactical echelons should be capable of gathering, analysing and using information obtained from contact with the population, security forces and the environment. If required, this human intelligence can be supplemented by technical intelligence, to the extent that resources permit. A second directive states: "Get out and walk – move mounted, work dismounted." Effectiveness grows from a better understanding of the environment; it is important to work in direct contact with the population, and thus to patrol on foot as much as possible.

In parallel, the idea of a "precision action" remains valid.[1] The effectiveness of the action in the human environment makes it necessary to avoid the uncertain actions and collateral damage which can all too easily ruin the painstaking task of winning minds. However, this sort of precision action can only be based on intelligence gathered by men or about men: the importance not only of HUMINT, but also of the actions of reconnaissance units and the widespread sensor network constituted by each and every soldier, is evident. "Every soldier a sensor" indeed. The current operational environment quite clearly highlights the role of the soldier in the intelligence process. The idea is steadily gaining ground that it is no longer enough to "imagine the enemy"; one must "think like the enemy". The enemy is neither inert nor a simple planning goal. Endowed with a free, creative spirit, he is a "subject" as well as an "object", and has no intention of thinking like us and adapting to our ideas. One must get inside the skin of the opponent. Following their difficulties in Iraq, the British and the Americans realised the need to create "red cells" in their operational staffs, and are now developing courses where one can learn to "think like the enemy".

We should be in no doubt that tomorrow's war will be conducted more successfully if it is seen as an intelligence operation, rather than a process of destruction. It will therefore be necessary to make not only an intellectual effort, but also human and material efforts, to promote intelligence, most probably leading to the provision of key assets at lower tactical levels.[2] We will most likely have to accept a reduced capacity for action, in

---

1. In fact, it is becoming even stronger, by acquiring a second dimension. Whereas in the past we thought in terms of "accuracy in space", we must now also think in terms of "accuracy in time", supplementing the CEP (circular error probability) with TEP (temporal error probability). This means that it must be increasingly possible for whoever is "tracking" the target to fire from the ground at precisely the right moment. It also means that firepower must always be available, at very short notice. There is an interesting example of this when, on the orders of US Defense Secretary Rumsfeld, the first American force (USMC) was sent into Afghanistan at the end of 2001, it took no artillery, in line with the new theory that air-to-ground fire alone would be able to deal with everything. The troops, often engaged in hand-to-hand combat in the hard fight against the Taliban, very quickly called for their mortars and howitzers to be sent from the US. The average delay in providing fire support was 20 minutes for close air support (CAS), as opposed to 4 minutes for an artillery weapon. 20 minutes can seem like an eternity when you are lying on the ground and under enemy fire.
2. A simple, but very clear example: the future structure of the US Future Combat Systems (FCS) brigades has been made public. In addition to an intelligence company and a reconnaissance and target acquisition battalion, each brigade will have a reconnaissance squadron for each of the joint groups. The other brigades, both light and heavy, will also have an organic intelligence company and a reconnaissance battalion. It is interesting to note that, during the 1990s, just when the manoeuvre areas were getting bigger and when uncertainty was becoming the opponent's primary characteristic, the French Army chose to do more or less do away with all its reconnaissance regiments.

favour of a revised and more complete reconnaissance and intelligence capability, which is the key to the proficient, targeted and skilled use of armed forces that is required for modern military effectiveness. Looking beyond the sterile debates on the respective roles of specialist and non-specialist forces, and the purely technical visions focused on equipment and networks, it is time to take an in-depth look at both our capabilities and our methods, in order to coordinate what will be needed to fight tomorrow's war and what we have available. The operations that make up tomorrow's war are primarily intelligence operations: this will oblige us to change our thinking.

Managing the negative effects of new information technology

In future wars, it will be important to be in control of the potentially negative side-effects of the progress in information technology. Theoretically, the more information is available, the better our knowledge should be; in practice, this is far from being the case. The old adage is regularly still shown to be valid: too much information kills information. In reality, the more information available, and the longer it takes to process it, the greater the risk of being unable to distinguish the relevant from the worthless, the important from the pointless or, quite simply, the true from the false. Certainty is far more a question of understanding than of facts. The proliferation of these facts requires a processing capacity suited to the need for analysis in the time available. The problem nowadays is far less a lack of intelligence and more an information overload, which is in danger of swamping the ability of the decision-makers to process and absorb the information. There exists a dialectic between time and information. In war, knowledge has a limited validity. The time taken to assemble the information needed to consolidate this knowledge tends to make it out of date, while at the same time the ever-increasing pace of operations reduces the time available for gathering and processing intelligence. The acceleration of the operational tempo increases the need for speed in making use of the intelligence, while the growing size of the areas of operation multiplies the need for information. The quest for intelligence resembles an unwinnable race against time.

Moreover, far from always making the situation easier to understand, the partial clearing of the fog may produce the optical effect of making the remaining layers even more dense. As the mass of available information grows, so the demand for information increases. As information arrives, new questions arise and new shadowy areas are created. The more accurate the information becomes, the more gaps seem to appear. The nearer it gets to the time when the decision must be taken, and the more new unknowns become apparent, the more likely it is that certainty will collapse. This residual intellectual uncertainty may therefore lead the decision-maker to delay his decision, in order to call on new intelligence sources and await the final information, allowing him to penetrate the new "foggy" areas revealed by the intelligence already obtained. With the uncontrolled quest for intelligence comes the risk of deciding too late, since the ideal window of opportunity for a decision is usually quite narrow; once enough information has been assembled, the time for action has often passed, as Clausewitz had already realised: "Waiting to be fully informed before deciding is equivalent to choosing the manoeuvre retroactively which grants the enemy complete freedom of action." This is an example of General MacArthur's "too late", the two words he said could be used to summarise the history of lost wars. The risk of wanting perfect intelligence is that the event itself takes the decision. Thus, however great the intelligence effort to reduce the degree of uncertainty, it will never be completely eliminated. Decisions will always be based on incomplete, inaccurate and even contradictory information. Waiting for certainty before acting inevitably means losing the initiative and missing out on favourable opportunities. Today more than ever before we must be able to decide "in the dark".[1]

The elements of information received are only of use if they are correctly organised and analysed, in order to be transformed, step by step, into knowledge that can be used in decision-making. However, this knowledge itself will only be fully effective if it is shared promptly among all the actors who need to have it. In this sense, lifting the fog of war depends just as much on information being correctly distributed as it does on it being correctly gathered. Efficient dissemination of intelligence is thus essential to its operational value. Nevertheless, here again lurks the danger of information overload affecting the executive level if the higher levels have not played their part. A lack of method or discernment means information may be disseminated too widely and too systematically, down to the lowest levels, without the necessary re-processing. The principle of the controlled sharing of information is clear: only information which is of direct use should be sent to a recipient, who should not be swamped by a flood of information liable to emphasise his uncertainty rather than reduce it. An actor who is weighed down by pointless details will see his capacity for processing saturated and, not knowing what he should retain and use, will remain in a state of uncertainty. In order to combat the profusion of information, and to be able to distinguish the essential from the rest, the key

---

1. This subject is discussed at length in Deciding in the Dark, Desportes.

question is to identify the useful information that should be sought, and how it is going to be used. To this end, the modern principle of real-time information sharing should be strictly monitored, if the negative side-effects are not to outweigh the expected gain in effectiveness.

"THINK GLOBALLY, ACT LOCALLY"[1]

The very idea of a "major battle" has almost become irrelevant, for a number of reasons. The first is that it belongs to a "major war", which, in one go, has become both irrelevant and unlikely. The second reason is that there is no longer a "major opponent" to be destroyed in battle. The probable opponent is no longer a single, large entity; this dispersion and the decentralisation of his actions have become essential to his survival and his ability to achieve his aims. The third reason is the economy of effort. The first consequence of the inevitable arrival in the "infosphere" of even the smallest incident is that, paradoxically, we no longer need an event of strategic importance to create the strategic effect. A limited tactical, or even micro-tactical, event may in itself have strategic repercussions; it is precisely this that enables an irregular opponent to resist, and even defeat, a much stronger adversary.

The new opponent has no formal army and so cannot operate at theatre level. Consequently, he does not present the sort of target which would allow us to make full use of our arsenals, which were designed to produce a strategic effect. We thus also become subject to the opponent's rules: in tomorrow's war, no act of force will ever be decisive. We can no longer dream of destroying the opponent in a massed attack. Since we cannot lure the opponent onto our battlefield, we must seek him out on his. We must match his attitudes –strategic offensive, theatre defensive, tactical offensive– with our own: our only solution is to adopt the same attitudes at every level, but with a better manoeuvre. Thus, all operations will be minor operations and only their amalgamation will produce the overall effect; they will be local with, in most cases, no tactical connection since, given the new opponent's structure, the systemic effect devised in the past as the key to success in war is unlikely to be appropriate. In the future, we can expect each engagement to stand alone in itself and in its intention; it will no longer be the operational effect that determines the multiplicity of actions, but rather the overall political plan which determines their relevance and cohesion. During the summer of 2006, Hezbollah adopted a decentralised organisation, thereby giving its combat groups a high degree

---

1. The golden rule for success, according to US Senators and Congressmen.

of autonomy. Each link in the defensive chain thus created represented a new dilemma for the IDF; faced with this decentralised approach, their concepts based on the notion of a "centre of gravity" became irrelevant. It was no longer possible to win a campaign by means of a decisive battle.

Furthermore, in the extensive theatres of future wars, situations are likely to be extremely variable in time and space. If one considers the very useful model of the "three blocks war" put forward by General Krulak (USMC), at all levels, from the operational to the micro-tactical, armed forces will be required to conduct three types of operations: coercion, security and support to populations. This will be the case for the force as a whole, as well as for the lowest levels. Even a company may be called upon to carry out tasks relating to these three areas. It will be necessary to resolve rapidly specific problems requiring different responses, tailored to each case. Local solutions, calculated correctly on the basis of detailed knowledge of the contexts and situations, and applied sensitively, will be more important than the global, long-lasting solutions, which are nevertheless still frequently considered to be the most effective model. The new operational landscape is emerging: it will consist of a multitude of widely varying actions, carried out at the lowest tactical levels. This is becoming increasingly clear as communication, and thus influence, once again seems to be the key to winning any likely future war. It is at local level, through the effects of the "oil stain", that the global effect is created.

The future war will be a succession of tactical events, in which the key decisions will be taken at subordinate levels. The political effect should be conceived centrally, in order to give a purpose to the multitude of small-scale actions, but execution can only be highly decentralised, carried out by small units familiar with local conditions and able to adapt, thanks to the autonomy they have been granted. We are a long way from yesterday's models: we need to think differently.

INFLUENCING MINDS

Communication first and foremost

The history of the 20th century and the abiding image of conflict between two strong parties have led to an image of warfare which, in fact, relates to only one of its many facets: a confrontation between arsenals in which the principal effect sought, directly or indirectly, was destruction or complete disruption. Action was aimed primarily at producing a technical effect, leading, it was assumed, to the political effect. There were many psychological effects pursued, but these were seen more as a means to an end –which was not the same– than as an end in themselves. Communication was centred on destruction, and the need to destroy, because this was felt to be the most important aspect of war.

The reality of future wars has gradually enabled us to return to a better understanding of the real role of war, which is communication: communication with an opposing power, with a population one is attempting to control and sometimes –for reasons of internal politics– even with national public opinion. Psychological dominance will be to the wars of the future what dominant terrain movements were for a long time to wars of the past. This reversal of roles between war and communication is even more flagrant, since destruction in itself seems to be less and less likely to lead to political effectiveness; the political success of communication through war would generally seem to be diametrically opposed to the traditional idea of a major military victory... however, it may sometimes be preferable not to respond and to allow the opponent to escape in order to prevent an increase in violence or destruction, or even to safeguard the existence of the indispensable partner in tomorrow's dialogue.

One might even say, that the shift from the model of industrial wars to that of war fought among the population has led to a fundamental reversal of roles: in the past communication was "about" war, while nowadays communication takes place "through" war. Military action has become a "means of communication"; any major operation is now a primarily a form of communication, in which all acts, even the smallest, speak louder than words. As we face our new opponent, who is skilled in using the power given to him by new communications technology, making good use of the power of images (faking organised massacres, torture, the execution of hostages, etc.) and attempting to exploit the media coverage of the reactions of the armed force, which he himself has provoked, our essential manoeuvre is communication. Conducting a war today, means managing the perceptions of all the actors involved, near or far, directly or indirectly. It is important to define the message to be passed on and to work out how it can reach the "global village", which nowadays encompasses Fallujah and Kandahar just as much as it does Moscow, Dallas, Liverpool or Paris. This means that war, far from being fought for its own sake, must be seen as a means of communication that must be included in an overall communication strategy consisting of many other facets. Initially, the use of armed force sometimes proves to be the most important; on the one hand, it is the most audible and, on the other hand, the initial armed scuffle is often essential in order to be able to impose the "strategic silence" in which the other forms of communication can be heard.

The prime importance of communication also means that modern armies have to be designed with this in mind and must have the technical capability to deliver the "right" message, which explains the need for accurate weapons, for example. They must also be able to be part of a long-term communication operation, in this case with the appropriate resources. They are destined to remain on the battlefield long after the heavy guns have fallen silent. The message and the way in which it is delivered, as actions speak louder than words, are fundamental in achieving the political aim. An individual action can now speak as loudly, as quickly and as far as the strategic plan and thus, since it is impossible to control everything down to the lowest levels, it is essential to ensure that the general intent is well understood by all, while leaving it up to each individual to adapt his own message to suit the circumstances.

In tomorrow's war we will not have to conquer, and even less to constrain, but rather to convince. This also means that we must think differently.

COMMUNICATING WITH WHOM?

In an inter-state war, the message of the war has only one target – the other state which we wish to convince to do our bidding. In a war taking place among the populations the picture is far less clear because, while it may be necessary to go through this initial phase –as illustrated by the war in Kosovo– the communication action is directed at multiple "recipients".

Communicating with the population

Naturally, the local population constitutes the first target for the communication operation, with the ultimate goal being to get the population "on side". Whatever their nature, any actions undertaken by the force are targeting the mental aspects even if, in order to do this, they have to involve the physical aspects; the way in which these actions are likely to be perceived should be an important consideration in their execution. An insight into the psychology of the populations among which the operations are taking place is thus essential, in order to understand their perceptions and anticipate their reactions. While communication may take place in many ways, the message is always the same; the foreign intervention is in the interest of the population and the political solution of which it is a part will lead to a situation which, overall, will be better than the previous situation or that sought by the opposing factions.

Of course, this message will only be audible if the "new situation" is close to what the population wants, and is not simply a matter of exporting other models. The mistake often made by the intervening power is to impose its own views and social, legal and political norms, and thus its own objectives, rather than trying to understand the needs and wishes of the population and ensuring the indispensable continuity demanded by history. One of the key difficulties encountered in recent interventions is the gulf between what the international community deems appropriate and the opinion of the local communities on these matters. The question of "standards" is very important; any answer can only be a compromise, a first step towards developments which will gradually get closer to the intervening power's idea of normality. In the early stages, at least, it is important to be able to tolerate certain aspects –even though to us they may appear shocking– of "their" normality, as they reflect centuries of traditions and social relationships. It is important also to remember that that humiliation imposed from outside is often a key factor in the eruption of political violence. Speaking on 10 October 2007, US Defense Secretary Gates[1] stated quite strongly that: "With the Iraqis, we have to establish relations in which 'shame and honour' play a greater role than 'hearts and minds'."

Moreover, the message sent will only be listened to if the sender does not seem too different or too far removed from the recipient. This raises the problem of the gap between the "intervener" and the "intervenee", and of the impact of one's own image, which is also altered by the difference in status between the party being imposed upon and the party imposing, if only in terms of shared values. In this sense, the physical barriers erected to ensure protection and the techno-centric Transformation of the Western forces do not make things easier. Thus, since the perception of the difference –the subjective difference– counts for more than the objective difference, there sometimes seems to be an insuperable gulf in understanding between the over-equipped forces and the underprivileged who occupy the crisis zones, by definition some of the most deprived areas on the planet.

At this point, it is worth taking a closer look at an expression which is often used without thinking about it. Are we really seeking, as is so often repeated, to "win hearts and minds"? While, undoubtedly, it is necessary to convince minds, in order to achieve the strategic objective, it seems that "winning hearts" is not necessarily such a good idea. Historically, it stems from a past which is very different from our current operations: in our colonial engagements the aim was to convince the local populations of a loving and giving home country, with values and virtues that they would be only too happy to serve. Today, the idea of a universal goodness borne by the intervening nations continues to justify this notion of "winning" hearts, in the same way as it justified the establishment of a filial link

---

1. Speech on10 October2007.

between those who were about to be colonised and the home country. While it is important to establish friendly relations with the local populations, this is seen today as a means rather than an end. The appeal should be made to their interests, sometimes global and collective, but more often individual and elementary, in order to allow the intervention force to operate; insurgency may have a political component, but it is only one aspect. The force can only win if it is better able to improve the circumstances of individuals than the opposing party. It is therefore necessary to convince everyone that it is in their interests to support the force, rather than the other side. Unfortunately, human nature, having been dissected by psychologists and sociologists, is all too familiar to us; with very few exceptions, and especially in difficult humanitarian circumstances, it is individual interests, linked to survival, that drive human behaviour. Since the logic of individual interests is the basis for collective interests, with each person trying to act in his own interest, it is important to meet individual needs in parallel with collective needs. In this sense, it is crucial to re-establish essential services as this represents a tangible sign of progress.

At this point, the sense behind the maxim rediscovered by the Americans in the strife in Iraq becomes clear: "Money is a weapon." Both for a crisis zone as a whole and at lower levels of action, money, and the ability to manage it sensibly, are frequently the most effective weapons in winning minds because of the concrete advantages they offer. Once combat ceases, considerable sums of money must be available and readily accessible: in order to ensure the desired political result, and thus the success of the intervention, the immediate views of the population are far more important than the more substantial, but more distant, effects of long-term investment. Thus we need to return to the logic and procedures long since abandoned by our army, in favour of a more intense style of combat.

It is necessary to act fast, as the level of frustration rises with the raising of hopes, but it is also important to ensure that only reasonable expectations are raised. According to Tocqueville's famous "law", improving the lot of all tends to increase, rather than reduce, general discontent, and the satisfaction felt by the individual depends less on the abundance of goods made available to him, and more on the ability of the collective to inspire in him desires consistent with what he can reasonably hope to achieve.

In other words, it is clearly the minds rather than the hearts that must be convinced of the interest of the foreign intervention. Over a hundred years ago, General Gallieni, spoke of "satisfying the needs of the populations, in order to persuade them to accept the new institutions". It goes without saying that legitimacy, already mentioned as one of the key criteria for the success of a mission, is fundamental in the battle for the minds. An intervention only becomes legitimate in the eyes of the outside world if it succeeds in improving the lot of the population, and in the eyes of the population itself only if it is able to fulfil the disappointed hopes which initially justified the decision to resort to arms to remedy the situation.

When communicating with the population, it is important not to aim at the wrong target, and thus to use the wrong ways and means. Experience shows that, broadly speaking, in the initial phase of an intervention the population can be divided into three unequal parts: 10 to 15% are in favour of the intervention, a further 10 to 15% are the adversary and the rest, about three quarters, are concerned only about their own individual interests, waiting to see how the situation develops before taking sides. For both the intervening force and the opponent, the real target is those who are sitting on the fence and who must be persuaded to come on side in order to secure a victory. This is the heart of the manoeuvre: to contain the opponent, to cause him to renounce his intentions or to destroy him if necessary, but first and foremost to use all available means –most of which are not military– to persuade this undecided mass that their interests, both individual and collective, will be best served by the success of the intervention. It is necessary to create symbols of true progress. Trust in the intervening force –in its impartiality and in its unselfish wish to help the populations involved– is absolutely key.

The decision by the undecided mass to join one of the sides acts as a rheostat regulating the influence of the two competing parties. The aim is to attract the majority of the population before the inevitable phenomenon of rejection has a chance to develop. To this end, the impression must be given that the intervening force is winning; the successions of small victories and the small improvements in the situation are signals indicating which party it would be best to follow. This is far removed from the idea of civil-military actions intended simply to promote the acceptance of the force; this initial stage is, indeed, essential, but understanding the paths to strategic success means going further and participating fully in the attempt to attract the population, without which the force would be condemned to carry out purely destructive operations, to no avail. It has become clear that the simplistic solution of "killing the bad guys" is no longer the recipe for strategic success.

It is worth noting that, to a certain extent, tomorrow's war has abandoned the dual model described by Clausewitz, to become a triangular struggle. Instead of fighting each other, the two parties are now more concerned to fight for the support of the majority of the population. The triangular nature of future wars –which, in effect, may simply be combining the two-way and three-way definitions of war as seen by Clausewitz–explains why the only successful strategies are those aimed primarily at gaining this support, rather than seeking to silence enemy weapons as a means to achieving an end.

The need to consider those sitting on the fence as the key to operations should not, however, obscure another facet of reality, which we have learned from history and experience: only someone who feels psychologically defeated will be prepared to abandon definitively the use of force and resign himself to the constraints imposed by the intervening power. The party that has been beaten must understand this fully and explicitly; in yielding to undeniable evidence, he must "accept" his defeat and not seek to find some new means of escape. This demand may mean that considerable losses are the condition for strategic effectiveness. Philippe Moreau-Defarges writes that: "For the intervention to succeed, the subject of the interference must be convinced that he has no option other than to submit and that he must play the game."[1] Without any doubt, the initial intervention must at least be sufficient to produce the psychological shock needed to change minds. After that, in certain situations, it may be enough to allow the crisis to continue towards its military conclusions, rather than stabilising an uncertain and inconclusive situation, since –no matter what those unaware of the reality of conflict may think– weapons can sometimes only be silenced once the cycle of violence has reached its natural end. It was because, in 1945, Germany and Japan felt truly defeated that these states could be reconstructed with no real opposition. It was for precisely the opposite reasons that, in 1940, France continued to fight at home and abroad, and that the people of Iraq, whose armies were not completely destroyed in April 2003, have pursued, in another form, the fight that they believed was not completely lost. In parallel with other measures, action must be directed at the adversary to convince him that victory through combat is impossible, and that his best interests lie in cooperating with the objectives of the intervention force.

Thus, as well as the need to conquer minds, it is also important to remember the need to convince the other party of his failure and of the futility of continuing his struggle. It is therefore necessary to be able, at all times, to use force –robust force– with all the technical capacity of Western forces. Theodore Roosevelt's fundamental idea is extremely relevant here: "Speak softly and carry a big stick." It is a question of credibility. In the past, this applied at the strategic level, the essential level of the action

---

1. *Droits d'ingérence* [The Rights of Intervention], Les presses de Sciences Po, Paris, 2005, p. 58.

at that time; today, it applies to the tactical level of future wars, where a display of force and determination is essential to ensure freedom of action, and thus success.

Communicating with the outside world

Progress in communications technology has created the second target for operational communication. Following the transformation of the world into a compassionate public space, each state has become part of a whole and is subordinate to this whole. The instant nature of communication and the opening up of societies have created a new continuity between the distant war and the judgement of public opinion, in which every spectator feels entitled to express an opinion, led by his emotions, which are rarely a good judge. In these conditions, tactical events –whether or not they have been exploited in this way– immediately take on a strategic and political dimension. Since each image can be used as an argument for or against, the second target for communication is the world outside the intervention; its importance grows as the legitimacy of an intervention is gradually established in world public opinion and among the populations undergoing the intervention. Tomorrow's war will be fought among the population – the presence of the media will ensure that it is more likely to be fought and won on our television screens than in the field. For this very reason, tomorrow's war will be a long war, as the role of the media will prevent the use of certain methods of combat that, in the past, produced rapid and decisive results. Regrettable perhaps, but true; once again, we will have to change our thinking.

While one of the difficulties regarding the media is to contain the excesses during the crisis and the early, very violent phases, another, in contrast, is to continue to hold their interest during the much longer, and much less gripping, reconstruction phase, often quickly overshadowed in "medialand" by another, more exciting, flare-up of violence. This challenge is fundamental for the strategic success of the intervention. The stakes involved in exiting the conflict are the key issue, but they only focus attention and political investment to the extent that the interest of the public –the voter and the taxpayer– is maintained. Communication is thus necessary in order to place the ups and downs of the stabilisation phase firmly on the media agenda.

Since the political effect sought in future wars is likely to be the re-establishment of a minimum state and normal living conditions for the population, military action alone will not suffice. During the initial intervention phase, its primary task is to create the conditions required for other actors to fulfil their roles. Thereafter, following the short period during which the residual lack of security makes it necessary for the armed force to carry out tasks which, under normal circumstances, would be carried out by others[1], its principal role will be, in various domains (security, humanitarian, economic, etc.), to facilitate actions leading to a return to normality.

Rediscovering the lines of operation

To an extent, we are observing the return to a phenomenon which somehow got lost during the confrontation between two blocs. Operations have always been the convergence of a certain number of "lines of operation". In this sense, one could even claim that the Second World War was finally won not on 7 May 1945, in Reims, nor in Tokyo Bay aboard the USS Missouri, on 2 September of that same year, but with the introduction of the Marshall Plan. However, during the Cold War, the dominant line of operation was indeed military; the vital nature of what was at stake reinforced its dominance to such an extent that all other aspects faded into the background.

Today, faced with our inability to achieve the desired political effects by purely military actions, we are rediscovering the irreplaceable role of the other lines of operation: diplomatic, economic, humanitarian, security, etc. The overall cohesion of actions in pursuit of a single political goal is seen, in theory, as the ideal for which we should be striving. The rediscovery of the "effects" of EBO will at least have been of significant pedagogical value; it will have transmitted the message that the purpose is more important than the capability and that a wide range of actors needs to be involved in order to achieve the desired political aim. It has also enabled us to realise that the "political" is more important than the "tactical" and, thus, that the long-term implications of actions should be taken into consideration in the same way as their immediate effects. For example, the destruction of networks, electricity generating facilities and vital infrastructure may speed up the tactical tempo for a while, but it will surely be detrimental in the long term if it slows down the strategic tempo, thereby delaying the return to normality and the eventual solution of the crisis.

Parallel lines

Recent operations have shown that actions along the different lines of operation cannot be conducted successively, as this would give the opponent advantages which he would not hesitate to exploit. Whatever prob-

---

1. With the exception of missions which would naturally be the province of the military, such as the reconstruction of local armed forces.

lems may be caused, achieving success in tomorrow's war means abandoning the idea of thinking and acting sequentially; effects must be coordinated simultaneously across multiple spaces, real and virtual, which have been rendered contiguous by modern communications technology. Any military decisions taken for purely technical reasons, as a result of initial insecurity, which neglect the political aspect, could have repercussions that could turn out to be very detrimental in the long term.

While security (combating delinquency and criminality, re-establishing and maintaining order, preventing violence) may appear essential, it should not be seen as a pre-requisite. Security is not achieved solely through security operations, but results from the convergence of diverse actions. Since the opponent often deliberately maintains a situation of security and instability in order to benefit from the war economy, and since irregular networks often have connections with criminal organisations, military action must unite with the police fight against organised crime. Today, the gradual restoration of the judicial system is considered to be essential, given the proliferation of organisations which see violent insurgency as a source of prosperity. When it comes to putting in place a minimum state apparatus, we now know that good governance is not the product of security; on the contrary, improvements in the level of security are just as much the result of good governance as vice versa. On the subject of the situation in Iraq, Larry Diamond commented that: "Good governance is impossible without at least a minimum level of security. But security cannot be improved without significant progress in the political domain."[1] The processes referred to as Disarmament, Demobilisation and Reinsertion (DDR) and Security System Reform (SSR) are neither "upstream" nor "downstream" processes, but simply processes which must be conducted in parallel with those of the other lines of operation.

Once the fighting has ceased, the force has no option other than to work steadily through the various, closely interconnected, lines of operations it has set itself. It is not possible to wait for a high level of security before making progress, perhaps more slowly, but in any case in parallel, in areas such as security, humanitarian aid or state-building[2]. On taking over the command of the US land forces in Iraq in January 2006, General Chiarelli, who had already had a lengthy and successful time there as head of

---

1. Larry Diamond, Building Democracy After Conflict – Lessons from Iraq, Journal of Democracy, vol. 16, no. 1, January 2005.
2. Of course, this is never an easy task, and it is difficult to establish trust; the fact that the concept of a state can be rather vague in the theatres of operation makes this even more true. This is the case in present-day Afghanistan, where the state has never really had much say outside of Kabul, and where the police have only ever been seen, at best, as a legalised organisation for theft and racketeering.

the 1st US Cavalry Division, warned his officers: "If you think that you have to wait for the area to be completely secure before attempting any reconstruction, none will ever get done." The first steps towards (material, social and political) reconstruction have to be taken even before hostilities have ceased, at the same time as the coercive actions. The force should not be looking for major tactical victories prior to undertaking the other actions, but rather for cumulative, modest and concrete advances along the various lines of operation, helping to move the campaign gradually towards its strategic success. The advances should be monitored closely, in order to allow results to be evaluated and directions and methods to be adjusted according to performance. It is important to note that, to fulfil this mission correctly, the force must have clear military superiority over its opponents.

Success, as has been said, assumes that the population is in favour of success and thus has rejected the contrasting option offered by the opposing factions. The support of the population is fundamental for the intervention forces, as well as for the insurgents; it acts as a sort of common "centre of gravity" for the actions of the two competing parties. Along the various lines of operation, this support and backing requires "positive" action –to add value to the intervention force's efforts– and "negative" action to counter the opponent and discredit his project. An incentive approach is preferable to coercion; the strategy for exiting the crisis cannot impose its ways and means on the population, as would be the case for public policies in a national context.

It is important not to do anything that would cause a physical or psychological split between the intervention force and its plan for the population. In particular, any law and order measure which terrorises or harms –physically or morally– the population will reinforce the position of the opponent, portraying him as the defender and enabling him to profit from the action-reprisal-retaliation cycle. On the contrary, efforts should be made to ensure the physical and moral separation of the population and the irregular forces, isolating them from their physical and moral support.

Future wars are likely to be long conflicts, in which spectacular and decisive operations will be the exception; their resolution will require a long-term, cohesive, inter-ministerial, inter-agency and often multinational approach. With its sights set firmly on the ultimate goal, any campaign must be seen as a whole, from preparation to withdrawal; it will need "backward planning", working back from the withdrawal and the desired end-state. Gallieni was an early supporter of unity of action, rejecting the distinction between political and military action. Unity of effort is a prerequisite for establishing the "decisive conditions" which are the major effect of operations in tomorrow's war.

Integrating actions

Interventions with purely negative aims –to end a massacre, attack, famine, civil war, etc.– rarely have a long-lasting, and thus useful, effect. It is therefore important that, from the very outset, an intervention goes beyond this stage and has a positive aim, in other words, one which seeks to tackle the causes of the problem. To this end, new situations must be created through the combination of multiple effects, and direct and indirect results of various actions. However, although it is necessary to advance along several lines of action simultaneously, these must all be part of a single operation, one which is truly integrated and designed as a whole in pursuit of an objective shared by both civilian and military actors. Any action on any line must support and facilitate actions on the others.

Integrating actions in this way is not without its own difficulties, as it involves merging widely differing logics. The case of NGOs operating in the humanitarian space is a familiar one, with the principles of charitable action often seemingly in direct opposition to political logic, while the humanitarian domain –with a long history of military intervention– is at the very basis of the legitimacy of state engagements. Whatever the problems, it is important to avoid creating a situation with, on the one side, those using weapons to communicate and, on the other, the "re-builders" using a different language. Support must be mutual and part of a united operation for the mission to succeed.

This integrated operation must be devised, synchronised and planned ahead of the intervention, in collaboration with all the actors involved. The problem in dividing up the tasks, however, is that it is based on a compromise between the unity of action and the sharing of the burden between the civilian and military partners; both are desirable, but are achieved at the expense of one or the other.

A relay race is an excellent image for civil-military cooperation. Well before the starting gun is fired, the participants have to study the race plan together. Then they need to train together. Once the race has started, without losing the momentum built up by the previous runner, they need to run together while passing the baton, an action which in itself is crucial.

Integrating actions in pursuit of a final vision

Crises, unrest and insurgency have a long gestation period. The success of the intervention is not linked to the cessation of armed confrontation, but to a sustainable solution to the crisis that led to the taking up of arms, and thus to the deep-seated underlying cause: effectively, the "decontamination" of the environment and the removal of the catalysts to the crisis. If the intervention is to lead to an enduring improvement to the situation, in addition to the direct imposition of minimal security conditions, it must also eradicate the roots of the conflict and create conditions likely to promote long-lasting, self-sustaining peace. Unlike humanitarian actions which, since they only deal with the consequences provide valuable, but only partial, solutions, tomorrow's wars will only be justified if they go further, since the ultimate question, and thus the ultimate response, are political.

The effectiveness of the integration implies the existence of an overall objective. It assumes that one knows not only what is needed "the day after", but the ultimate purpose. Thus, clearly, the intervention should be planned on the basis not of an initial tactical victory, but of the conditions for its strategic success.

As we now know, this was the problem with the plan for Operation Iraqi Freedom, devised as a classic battle, from the race to Baghdad to the fall of the Ba'ath regime. Neither in Washington nor in Tampa –headquarters of the US Central Command (CENTCOM), which was in charge of the operation– were the conditions of success after the battle taken into consideration. There was a rapid proliferation of looting and outbursts of violence, first in Baghdad, later across the whole of Iraq, a forerunner to the insurgency that followed. The "irregulars" were able to seize the initiative swiftly, as there was no plan other than the "strategic success as an immediate result of the tactical success". There really was no other plan than the spontaneous emergence of democracy: no "Plan B", no planning for anything other than the realisation of an improbable dream born of great strength and being out of touch with reality. Following a tradition which is well anchored in the American sub-conscious, the war was conducted as a technical action, separated from its political nature. This means that the military action –the technical factor– must be integrated into the other success factors. It is this integration, in view of a single aim, which gives the intervention its essential political substance.

It is only possible to manage the end of the intervention phase correctly –the exit from the crisis to which it is so closely linked– if this has been considered carefully at an early stage. Exit strategies must be set out prior to the intervention, and should involve all those taking part, public or private. American think-tanks believe that the most important lesson to be learned from the difficulties encountered in Iraq and Afghanistan is that early planning is crucial to success[1]; in both cases the plans were

---

1. Infrastructure Reconstruction: Imperative in the National Interest, Final report, May 17-18 2006, Washington D.C., p. 4.

drawn up and formulated only after the initial intervention and –particularly in the case of Afghanistan– were difficult to implement because of a lack of local actors suitably qualified to play an active role in the operations. The forces, and the other actors, must be able to seize any opportunities immediately and therefore must have the necessary resources available from the outset, for use as soon as the initial fighting has ceased.

One of the major problems with the intervention in Afghanistan (October 2001), was the fact that it was initially envisaged as retaliation for the attacks on 11 September, evolving only later into a "nation building" mission.[1] Since, at the outset, the sole objective of the mission was to destroy the Taliban regime, it took no account of the repercussions of the voluntary reinforcement organised by the "warlords", while the decision not to deploy ground troops removed all control of the theatre from the US once Kabul had fallen. In contrast to sensible practice, the war was not conducted with a view to the improved state of peace expected later, but rather solely with a view to the immediate result. The desire not to deploy troops on the ground from the outset has proved to be a very expensive one. As the head of the CIA, George Tenet, put it on 10 October 2001: "They have placed their destiny in the hands of Afghan tribes, who are going to act, where, when and at the speed that suits them. These tribes have their own problems, their own ambitions and their own internecine strife. They are a mercenary force which is not controlled by the United States. This is the price that must be paid for the initial decision that the tribes would fight on the ground, rather than the US Army."[2] The initial intervention in October 2001 did not destroy the forces of Al Qaeda, nor did it enable the capture of Bin Laden and his lieutenants; having started badly, the war in Afghanistan continues to encounter problems, resulting in an escalation of resources with no guarantee of success.

Early success

The first days following the initial combat phase –the so-called Golden Hour– are crucial. Maslow's pyramid gives a good appreciation of the priorities for the action (physiological needs, safety needs, social needs, etc.). No action will succeed until elementary human needs have been met; it is

---

1. It is interesting to note that, given the time it has taken to build their own nations, Europeans prefer the expression "state building" to "nation building", while Americans use only the latter. The difference comes primarily from the fact that, for the United States, state and nation historically form a single unit, indeed for the US, the state preceded the nation. This cultural difference also partly explains the American tendency to try first to rebuild the institutions, assuming that the nation will take shape around them.
2. Woodward, Bush at War, p. 193.

therefore important to respond first of all to the immediate concerns. The fall of a regime, especially a dictatorial regime, inevitably leads to a decline in public order, which must be countered immediately before the ensuing chaos destroys all hopes for the future or violent factions take charge. Looking at the situation in Baghdad at the end of April 2003, John F. Burns wrote: "Almost as much as the civilian casualties from American bombs and tanks, the destruction of the museum and the library has ignited passions against American troops, for their failure to intervene. [...] The cure has proven worse than the disease – having many of the city's principal institutions laid to waste by looters has been too high a price for freedom."[1] It is thus essential to be able to conduct, without immediately seeking continuity, actions combining both civil and military in order to rebuild the "social contract" and establish acceptable living conditions for the population as soon as possible.

Ensuring the means for integration

This key principle of the unity of effort can only be applied if an inter-ministerial process for identifying needs, evaluating capabilities, allocating responsibilities and, thus, planning actions has been established before the start of the operations. The concept of "integrated operations" from the outset should be a guiding principle during the pre-operation planning phase: the operation should be planned "in reverse" on the basis of the desired new state and the difficulties that are likely to be encountered. A lack of coordination will prolong the military deployment –which can swiftly become counter-productive– as it slows down social, political and economic reconstruction. This essential early integration requires the development of structures facilitating a multilateral and inter-ministerial approach, both initially and in the long term.

At the same time, and perhaps more importantly, it is also a question of culture, which must be adopted by the military. The military must understand that areas which, in the past, were not part of their usual thinking and operations, have now also become "their business", part of their job. No longer a matter for experts, who are seconded to military structures as required, these areas are now part of a general know-how and attitude. It means that military units will integrate into their organic structure not only experts in operational communications but also specialists in Civil Affairs (economic reconstruction, education, information, governance, infrastructure, application of the law, health, etc.).

---

1. John F. Burns, Baghdad Residents begin a Long Climb to an Ordered City, New York Times, 14 April 2003.

UNITING INTERESTS AND RESPONSIBILITIES

In societies emerging from a conflict, security, economic growth and the establishment of a minimum state are interdependent. Economic growth is a vital factor in preventing the return of violence; it is therefore essential to create a situation in which people, goods and services can move freely. In addition, the cost of "reconstruction" can be high. It is important to seek a means of sharing this among all the actors who have an interest in the success of the reconstruction and whose involvement can help to achieve the desired result.

Involving local actors

The first people to be integrated into the joint effort are, of course, the local actors. Since success can be measured in terms of a self-sustaining peace, it implies ownership by the local population, achieved through their gradual involvement in the process. Rather than following the destruction with a period of imposition, which could, bizarrely, result in the population becoming dependent on the intervention force, reconstruction should be a joint undertaking, in which the population are actors, not just spectators. It is important to interest the majority in the success of the operations, in support of the actions agreed by the intervention force and the local leaders.

Recent crises have shown us the futility of attempting to change a political system from outside and how difficult it is to reconstruct a social contract, reconstitute an unwilling nation, or to build a state on imported ideas. It is also hard, and possibly even counter-productive, to demand that the other side renounce what he stands for, by refusing to accept his specific characteristics, even if they do clash with our ethnocentric idealism. On the subject of involving local populations and respecting other cultures, the words of Lawrence of Arabia may have a familiar ring: "Do not try to do too much with your own hands. Better the Arabs do it tolerably well than that you do it perfectly. It is their war and you are to help them, not to win it for them. [...] The less apparent your interferences, the more your influence." In order to respect this perennial truth, the only option is to make use of existing structures and elite groups; both are the product of a long history, which no rapid action can replace. Lyautey puts it clearly: "We must govern with the mandarin and not against the mandarin [...] in every society there is a ruling class without whom one can do nothing."[1]

---

1. Lyautey, *Lettres du Tonkin et de Madagascar 1894-1899* [Letters from Tonkin and Madagascar 1894-1899], 2nd edition, Paris, A Colin, 1921.

In a similar way, sixty years later, in 1953, a note from the French High Command in Indochina, setting out the principles of pacification in Vietnam, stated: "Pacification is a purely Vietnamese national matter. Wherever possible, the military command should therefore step aside in favour of the national administration."[1] US Defense Secretary Gates said exactly the same thing a further five decades later: "The most important military component in the War on Terror is not the fighting we do ourselves, but how well we enable and empower our partners to defend and govern their own countries. The standing up and mentoring of indigenous armies and police –once the province of Special Forces– is now a key mission for the military as a whole."[2]

The involvement of local actors will make it possible to relaunch the local economy and, at the same time, to provide employment and financial income which will obviate the need to side with the opposing factions for help and support. The adult population needs something to occupy them other than violence, but also needs to have something to live on. This is a difficult challenge, since the irregular opponent can frequently provide a source of income quite out of proportion to what can be offered by the projects backed by the intervention force, at least in the short term.

The relaunch of the micro-economy should make it possible to sever the negative connection between a lack of employment and irregular combat: the local antagonists have no work because they are fighting, but they also fight us as long as they have no work. Some form of non-violent competition must be established between the various parties. There is, in both senses, a direct link between restoring security and economic growth. Reconstruction itself has a tactical effect, since it reduces inactivity and thus insecurity. Restoring local economic activity should not be seen as simply improving living conditions; it is an important part of re-establishing administrative and political stability. Just as in our own societies, in the theatres of operation there is a strong link between prosperity and security.

Furthermore, success cannot be founded on disunity and dissension; on the contrary, it requires the resolution of the political and social contradictions at the root of the crisis. Integration will thus be far more important in solving the crisis than exclusion. In tomorrow's war, defeating the opponent will have no meaning if it does not lead to responsibilities ultimately being transferred to a solid, local authority, accepted by an overwhelming majority of the population. As soon as the intervention force arrives, it

---

1. Note 92/PAC/CVN of the *Bureau Régional de Liaison pour la Pacification* [Regional Pacification Liaison Office] (SHAT-10H 3167).
2. Speech on 10 October 2007.

becomes an instrument of social and political engineering, willingly or unwillingly. From the outset, it must therefore be prepared to share in order to involve, and to involve in order to attract and to rebuild freedom "together", as freedom cannot simply be claimed. Such sharing must be approached gently. It may seem odd, but justice and political stabilisation do not necessarily make natural bed-fellows. Purely for reasons of effectiveness, in these new contexts, it is wise to remember Churchill's famous maxim: "Magnanimity in victory." This is particularly true in view of the importance of force ratios on the terrain; we have no say in choosing either the actors or the discussion partners, whom we must nevertheless bring together to work towards a common aim, in a sort of "peace of the brave".

It is a truth universally acknowledged that the decision taken in the summer of 2003 by the Governor of Iraq, Paul Bremer –as part of the process to rebuild a democratic regime from scratch– to dissolve the army and security forces and to remove all members of the Ba'ath party, at whatever level, from the ministerial structures, was a grave error. He decreed that in the short term there would be no Iraqi government, thereby introducing a de facto period of occupation. Thus the country found itself in a disorganised state, with no national leaders and no organisations, other than religious ones, and with hundreds of thousands of civil servants and other workers now at the bottom of Maslow's pyramid and prepared to make their skills and their weapons available to the burgeoning insurgency. It is interesting to note that the US military hierarchy in Iraq was opposed to this decision and that, according to General Peter Pace, Chairman of the Joint Chiefs of Staff, the decision was taken without consulting the Pentagon.[1]

An essential regional perspective

An intervention can only really succeed if it is conducted with a regional perspective. It is essential to have a full understanding of the impact of neighbouring states or ethnic groups, cross-border traffic, the creation of safe areas or infiltration by violent factions.

In reality, neighbouring states often have a political interest in the resolution of crises, whose effects always extend beyond the actual theatre of operations; they may also be keen to exploit economic interests In any case, it is important for the powers behind the intervention to set goals which are compatible with the interests of the neighbours, since experience shows that it is virtually impossible to resolve a crisis if neighbouring states are out to wreck the efforts. The only solution is therefore to

---

1. International Herald Tribune, 5 September 2007, p. 4.

develop a constructive relationship, notwithstanding any previous role they may have played in the crisis.

DEFINING THE ACTIVITIES OF PRIVATE MILITARY PROFESSIONALS

Private military companies (PMC) have developed rapidly, to the extent that they are now a major player in modern conflicts. The war in Iraq highlighted this phenomenon, which had been growing steadily for some time. It had already increased as a result, on the one hand, of the radical reduction in the size of the US Army following the end of the Cold War and, on the other hand, of the decision by Defense Secretary Rumsfeld to engage resolutely in the rationalisation of the military instrument, with a corresponding transfer to the civil sector of skills judged not to be strictly military in nature. In the field, these auxiliary troops often outnumber regular troops; in mid-2007 the US Department of Defense employed 185,000 subcontractors in Iraq, as against 160,000 regular soldiers. Armed forces rapidly discovered that these PMCs offered considerable advantages, both economic and in terms of flexibility. States gain freedom of action and increase their capability.

There is a strong, some may say inevitable, trend here, which must be kept under control, since by no means all its effects are positive. For many people, this represents a weakening of the state, causing it to lose its supreme right to the use and control of violence; it is a move away from Max Weber's classic vision, in which "the state has a monopoly on the legitimate use of violence". In 2006, the Chairman of Blackwater USA announced that his company had available, "a brigade-sized force ready to deploy in crisis zones at short notice".[1]

A study of the cultural background to this movement makes it easier to understand the risks. While there is no doubt that mercenaries have existed as long as war itself, the present model is British and American in origin, but primarily American. As early as 1965, Business Week considered that the Vietnam War was a "war by contract".[2] This opinion was based on a number of factors, one of which was the process of re-training US military personnel to provide an unbroken link between the public and private sectors and between civilian and military life. This produced a convergence of interests built up throughout a military career. The American system of "contractors" also played a role; the US armed forces have no agency equivalent to the French *Délégation Générale de l'Armement*

---

1. The Virginian Pilot, 30 March 2006.
2. Vietnam: How Business Fights the "War by Contracts", Business Week, 5 March 1965.

(DGA) [General Delegation for Ordnance], but instead contract out directly to private companies. This practice favoured the growth of outsourcing when, immediately following 11 September 2001, there was a dramatic increase in military budgets. Another key factor is the convergence of interests of the PMCs and major American industrial groups, which have found new outlets for their products. Most PMCs, including the largest (MPRI, Dyn-Corp, Vinnel Corporation, Halliburton, etc.) have now been bought up by major industrial groups, leading to a veritable spate of mergers and/or acquisitions since 2001.

However, in the United States, turning to private companies in this way should be seen primarily as a means of avoiding control by Congress and American public opinion. With regard to the deployment of armed forces outside the United States, it offers a way of overcoming the difficulties experienced by the government in involving armed forces in peripheral conflicts and in justifying casualties in engagements which do not appear to be directly linked to vital interests. In stark contrast, the four corpses, lynched, burnt alive and strung up on the bridge at Fallujah on 31 May 2004, were rapidly forgotten, even though this incident did later lead the American forces to launch a bloody response. After all, these dead civilians were only Blackwater employees, just four victims amongst the 1000 PMC employees killed and the 12,000 wounded in the four years of fighting since the fall of Baghdad.[1] The use of PMCs also makes it possible to avoid restrictions placed on the use of public finds. Despite the dramatic increase in the defence budget in the wake of the attacks of 11 September 2001, allowing a considerable effort to be made in terms of equipment, the money available for personnel remains inadequate. Thus, private military companies provide a way of making up for this lack of human resources through contracting out, without the need to beg for Congressional approval. In addition, they act as a means of artificially compensating for the difficulties in recruiting, given the increasingly poor light in which the current conflict is viewed. This method made it possible to overcome the restrictions imposed and to artificially reduce the number of military personnel engaged in operations, while the lack of transparency in the contracts favoured the use of PMCs in clandestine operations. In the case of the USA, one can say without exaggeration that the widespread use of PMCs is an instrument of military freedom of action which can only be seen as contrary to the democratic tradition of political control.

In addition to this fundamental problem, this policy of privatising war raises a number of other questions. It involves a number of risks, only

---

1. New York Times, 19 May 2007.

some of which are discussed here. In the background are two questions relating firstly, to the status to be given to these people engaged in military operations (Regular troops or mercenaries? What is the legal position in the event of capture?...) and secondly, to the responsibility towards the employee and his activities (Who is responsible, the state or the company? Under whose jurisdiction does he fall?...)

More concretely, there is the question of the safety of the personnel. The protection of civilians involved in combat operations poses serious problems; it may require the use of considerable assets, thereby negating the fundamental principle of making savings by employing PMCs. The very loose control over this resource and its use disrupts meticulous military planning. The risk of an ill-timed withdrawal is only one of many. There have already been occasions where a deteriorating security situation has led private military companies to withdraw abruptly from theatres of operation. The US Army has also frequently found itself either caught up in operations resulting from incidents linked to the uncontrolled activities of PMCs, or having to take on missions it had previously delegated. Furthermore, despite the actual reduction in the cost of salaries, social security contributions, allowances and pensions, in the end the financial dimension seems more of a rationalisation than an accounting reality, as the true cost is often higher than predicted because various associated costs, such as the cost of protection, were not included. In any case, the large number of contracts, together with the web of companies that signed them, makes it difficult to calculate the overall cost. It is also well-known that the rise in the number of PMCs was immediately reflected in a loss of skills, acquired at high cost, due to the significant departures suffered by elite American units in particular, but also among Western forces in general.

There is also the question of the interests of the companies present in the field. Most are transnational companies, whose interests do not naturally mirror those of the states employing them; only rarely is there any idea of "national interests". Nor is there any evidence to show that the PMCs employed are in favour of a rapid solution to the crises which form their source of revenue.

In addition to the risks involved, the behaviour of PMC employees is a cause for concern. As General Nash put it: "If you're trying to win hearts and minds and the contractor is driving 90 miles per hour through the streets and running over kids, that's not helping the image of the American army. The Iraqis aren't going to distinguish between a contractor and a soldier."[1] The ethics of military sub-contractors are decidedly uncertain

---

1. Washington Post, 5 December 2006.

as, generally speaking, their recruiting is neither monitored nor controlled by the state. It is worth recalling that all the interpreters, and almost half of the interrogators, involved in the Abu Ghraib prison scandal were private contractors employed by Titan and Caci but, in the absence of applicable legislation[1], these individuals could not be brought before a court. The populations in theatre, quite rightly, associate these contractors with the state, which is thus likely to see the all-important legitimacy of its action eroded. In a theatre of operations, it is increasingly difficult to distinguish between the various actors present: regular forces, special forces, who sometimes wear rather "exotic" uniforms, and civilians employed by private companies. It was Nietzsche who cautioned: "He who fights with monsters might take care lest he thereby become a *monster.*"

The most dangerous aspect of all this, is that outsourcing accentuates the trend towards the weakening of states, at a time when increased globalisation should really be counterbalanced by an anchoring of certain prerogatives belonging to a state, and when, paradoxically, the very aim of Western intervention is "state building". Since the Treaty of Westphalia, the monopoly and control of the use of violence had become the major foundation of the nation-state; the weakening of this principle undermines the model which guarantees at least a certain degree of world stability. Accepting that the sovereign power constituted by the right to wage war can be delegated means accepting the weakening of state power through the creation of transnational powers. Charles de Gaulle warned us of this as long ago as 1952, in a speech in Bayeux on 14 June: "Defence is the primary raison d'être of the State; it cannot neglect it without destroying itself."

Today, it would be as unrealistic to reject all forms of privatisation of force, as it would to claim that it is possible to exercise tight control over such privatisation. However, it would be irresponsible to refuse to acknowledge that there are aspects which directly damage the image of the Western world and of the values that support the legitimacy of our interventions. It is precisely this image and these values which legitimise our intervention and thus, in effect, our ability to go out and ensure the security of France from outside our national territory, something which is becoming increasingly necessary. In other words, if we wish to continue to

---

1. However, under the terms of an amendment to the Defense Authorization Act 2007, proposed by Republican Senator Lindsay Graham, and adopted in October 2006, civilian contractors are now subject to the code of military justice in the case of a "declared war", as well as in the case of "an emergency operation". This provision should be extended to civilian contractors employed by ministries other than the Department of Defense.

benefit from privatisation, while curbing its excesses, it is important to keep a tight rein on the activity of these private military professionals. Following the initial euphoria, in response to the daily abuses and blunders, this new idea is taking shape in the United States, and in Congress in particular: American legislators are keen to ensure a much tighter control of these companies.

Following fashion can be dangerous. Experience in other countries shows us that we should avoid being seduced by chimeras offering the opportunity to increase numbers which, in the end, prove to be more costly.

CHAPTER IV

MAKING THE BEST USE
OF TECHNOLOGY

Technological superiority is not an end in itself. It will not be sufficient to solve the problem of war.

However, on both sides of the Iron Curtain the dialectic circumvention of the Cold War primarily involved the use of technology, which eventually acquired an autonomy that still exists today, even though the initial logic behind this development has completely vanished. The autonomous development of technology has lead to what could be called "technologism", in other words, a blinkered attitude, which can make it impossible to take a step back and consider whether today's answers do indeed correspond to yesterday's questions. Caution is required, if we are to ensure that our staffs do not succumb to the danger of concentrating on "a strategy of means", without asking "what means?". Particularly in the field of equipment, the end-state approach should be given precedence over the capacity-based approach, which soon becomes irrelevant if it is not part of a political vision.

Armaments should be seen as a function of their military effects, but not exclusively so. The key aspect is their ability to play a useful role in achieving the desired political effect. It is therefore important to assess their direct and indirect effects in areas very different from pure military technology. Once again, the political substance of arms has become the essential factor.

Technology, particularly very high level technology, is indispensable. We just have to look at it differently.

RETHINKING THE ROLE OF TECHNOLOGY

Circumventing the strength of the other side by means of technology was one of the most characteristic aspects of the opposition of East and West during the Cold War. The cause disappeared, but the process has taken on a life of its own, supported, it is true, by a range of interests. During his farewell speech, on 17 January 1961, that great soldier, President Eisenhower, already warned of the negative effects of this burgeoning logic, which had become detached from its initial rationale.

By the mid-1990s, the absence of an enemy meant that technology had to find a new official rationale. This turned out to be the "capability-based approach", strongly supported by Defense Secretary Rumsfeld as soon as he took up his job in 2001: since we do not know who tomorrow's enemy will be, we need to guard against all threats and surprises, and move from a "threat-based approach" to a "capability-based approach", intended to dissuade any adversary from taking on the United States. The only solution open to the US was to stay ahead in all areas, in other words, to consolidate its position as the only superpower, and to remain stronger and more invulnerable, making it the best in all fields of military action. Thus technology became what amounted to a rejection of the true nature of war –a dialectic of wills– in that it gave the political decision-maker the ability to escape from a decidedly outclassed potential opponent. Technology became an end in itself and a substitute for strategy while, as the iconoclastic analyst Ralph Peters, put it: "In this age of technological miracles, our military needs to study mankind." To see war purely in terms of technology is, of course, quite wrong, since it is social and political factors that define it.

The concept of circumvention by means of the highest level of technology no longer makes strategic sense. As a result of the proven decrease in the operational return of technology, it is necessary to redefine the balance between the manpower strength needed for modern crisis management and the amount to be invested in technological support, which is always needed, even if "asymmetric cunning" –which, through clever use of the population, is contemptuous of "sensors" and "shooters"– negates many aspects. The attempt to get performance for performance's sake is the reason for the staggering costs of major weapon systems; since the beginning of the Second World War the cost of a main battle tank has increased over three hundred-fold, and this ratio can be multiplied by ten for fighter aircraft. Here, too, we will need to think differently and adopt the policy of having just the right level of technology, and of the right technology, by finding a balance between the level of technology we are prepared to fund and the size of the armed forces we would like to have. This recalls the "law" announced in 1978 by Norman R. Augustine, who, prior to leaving his senior position in the Pentagon to be head of Martin-Marietta, and later Lockheed-Martin, observed the extraordinary rise in the cost of military aircraft: "By 2050 the entire defence budget will be enough to buy a single tactical fighter, which can be used three days a week by the USAF, three days by the Navy and the last day by the Marine Corps". This exponential increase in costs is diametrically opposed to those routes which indicate uncertainty about the future and the speed of technological change; the greater the choice of weapons, the more expensive they are, and the less room there to develop and adapt. Ralph Peters again: "We are seduced by what we can do; our enemies focus on what they must do."[1]

Only immoderation is reprehensible. Of course, if the Western powers were not to retain their conventional fighting capabilities, we would naturally return –through the same phenomenon of circumvention intrinsically linked to strategy– to this form of war. Thus the only way of avoiding conventional confrontation is to retain these assets, in the same way that there is no alternative to retaining our nuclear arsenals, at just the right level. The difficulty facing us today is the need to have a wide range of capabilities. We need an army that is able to operate both at home and abroad to ensure security and to safeguard the population, but it must also be able to perform a range of other tasks since, unlike the situation during the Cold War, there is now an infinitely variable array of possible engagement scenarios and adversaries. However, all Western armed forces are subject to budgetary constraints and, if we do not take care, the great rise in the costs of certain armaments will swallow up the available resources, at the expense of assets considered, wrongly, to be of secondary importance while we are only dealing with a virtual war.

The choices are not easy, but a clear path seems to be emerging: once we have reached the point where, whatever the interested alarmists say, in an open area at least, no army in the world is able to withstand a Western coalition, should we not restrict the trend towards ever more sophisticated technology, which is diverting military budgets from their prime purpose? In other words, should we carry on with the general strategy steered by technology, or should we install this alternative strategy in its place? In the light of the new requirements emerging from the developments in any likely future war, the relevance of the increase in this technological differential is coming under close scrutiny. Other needs and other urgent

---

1. Weekly Standard, 6 February 2006.

requirements are emerging, in particular the urgent need for "control". This requirement became virtually obsolete in the confrontation models of the Cold War, but the control of the battlespace and of the surrounding areas has once again become an essential part of establishing the conditions for strategic success; it has replaced the need to destroy, which was the prime condition during the Cold War. Contrary to what some people continue to affirm –despite the evidence– fifteen years of "future engagements" have demonstrated that technology can only partly resolve the question of numbers. We still need to have sizeable forces available to us. The defence and protection of France depend on our ability to achieve stabilisation "at the front", and the problems of stabilisation can be solved only by a clear, and lasting, presence on the ground. Tomorrow's war will be based on control, and control implies numbers in our ever-expanding areas of operation.

The reason the United States is having problems in Iraq, despite the swift capture of Baghdad in 2003, is probably because –ignoring the recommendations of US generals– it did not have enough troops on the ground from the start of the intervention and has thus been unable to exercise "control" since the end of the initial, very short classic phase. It seems to be very difficult to make up for this initial error, just as it was in Afghanistan in 2001, and as it was for the Israelis in July 2006. It has now been firmly established by current and past operational experience that quantity has once again become an important quality; thus, deliberately exchanging high cost, very high technology for numbers with the right level of technology, equates to exchanging technological obsession in favour of the effectiveness and utility of the force. It also means endangering the lives of troops deployed in insufficient numbers.

European nations must resist an American principle which suits neither their budgets nor the new operational effectiveness: "Everything that science has to offer is worth trying", as Ambassador Andréani put it.[1]

We should follow the converse principle: "Not everything that is technically possibly is necessarily desirable."

The future does not lie in technology; the future lies with mankind. Technology simply offers us a means of dealing with the future. The history of war is not the history of technology, but rather the history of human beings. Technological temptation has been around for a very long time, particularly in military thinking which, by nature, is always seeking to be more effective and to improve the tools available. Over forty years ago, General Beaufre condemned "this attitude of realistic appearance, which

---

1. Andréani, p. 24.

sees strategists as backward and pretentious and which concentrates instead on tactics and materiel, at the very moment that the speed of change requires a high-level overall vision."[1] However, observing the effects of technology in the settling of disputes should give cause to doubt its ability to solve all the problems of war, which are fundamentally insep-arable from human nature. Technological superiority did not prevent the Dutch being defeated in Indonesia, nor the French in Indochina or the United States in Vietnam nor, even more recently, the Soviet Union in Afghanistan or the difficulties encountered by Russia in Georgia and Chechnya. We should be cautious: technology offers no absolute guaran-tee of superiority, either on the battlefield or at the negotiating table. As Jomini observed over two hundred years ago: "The superiority of arma-ment may increase the chances of success in war; it does not, of itself, win battles." This is echoed by the war conducted by Israel in 2006, which reminds us that the way in which technology is used and the skills of the armed forces and the men are far more important than the technology itself.

Technological obsession is not a new phenomenon, even though sol-diers, strategists and politicians have always known that technological prowess and strategic competence are completely unconnected. For over two centuries, each era has been affected by the upheaval caused by the arrival of new techniques. This idea is wrong: the upheaval comes not from the technology, but rather from the spirit which masters it and makes the best use of it. Success does not come from technology, but from the asym-metric idea that uses it to its advantage. Sadly, the obsession with technol-ogy usually takes precedence over the idea: all too frequently, as Colonel de Gaulle wrote in his memorandum of 26 January 1940, "we simply inte-grate into the already established system, the new means offered by the era". When incorporated into old organisations, and used according to outdated concepts, new technology frequently offers only marginal bene-fits. In 1870, the French Army's machine guns could have at least slowed down the Prussian forces, who had none, but having been left behind in the artillery batteries, they were of very little use. Liddell Hart and the future head of Free France agreed that restoring mobility on the battlefield was not simply a matter of introducing tanks and armoured vehicles. It also required a revolution in military thinking, for example that prepared by Von Seeckt and implemented by Guderian: "The Germans won not because they had more tanks, but because they knew better how to use them."[2] In his writings, Foch reminds us that what counts is not so much

---

1. Beaufre, p. 31.
2. Coutau-Bégarie, p. 252.

the latest technology, but rather the crucial importance of integrating it into revised organisations and concepts, where it can be used to its best effect; at an early stage, the choice of technology must be based primarily on concepts and strategy. The fact that the Maginot Line, which brought together the most modern technology of the day, consumed the greater part of the available budget counted for nothing; it simply led to defeat. The problem was that it had been conceived according to the models of previous wars, and simply enhanced with new capabilities, forgetting two of the most basic rules of war: circumvention and the intelligence of the opponent.

Inherent in this obsession with technology is the risk of confusing war with the tools of war, that is to say, of overestimating the role of arms and forgetting the intangible aspects of war, the central elements obscured by the events of the day. Colonel Ardant du Picq called for greater prudence: "The new weapons are of almost no worth in the hands of weak soldiers, no matter how numerous they may be." At the same time, on the other side of the Atlantic, Alfred Thayer Mahan wrote in one of his books on the influence of sea power that: "History teaches us that good sailors in poor ships and better than poor sailors in good ships." Technological advantages may waver, but the basics of war remain. What experience shows us is that for equipment, what counts is not its technological potential, but rather its ability, in the probable conditions of engagement, to play a role in ultimately achieving the political goal.

TECHNOLOGY NO LONGER GUARANTEES SUPERIORITY

Technology itself has very rarely decided the outcome of a confrontation, as it is only one element of strategic effectiveness; indeed, its influence is often very much less than the attention given to it. While every technological advance helps to clear the fog of war a little, it also introduces its own new frictions and causes its own errors. However, war is primarily a social phenomenon; apart from the specific technical advantages offered by the quality of the tools (the weapon systems), the outcome of a confrontation is primarily dependent on the environment –political, economic, cultural, geostrategic– of the two sides. Similarly, a technological imbalance is a far less important factor in strategic success or failure than a sound political choice, strategic or tactical, which usually makes it possible to compensate for technical inferiority. Moreover, in tomorrow's war, among the population, our best technology will often be found wanting simply because it will be unsuitable for the circumstances; our opponent, aware of our advantages will find ways to circumvent them and render them useless.[1] Sir Rupert Smith, for two years General Officer Commanding Northern Ireland, observed that the IRA always made sure that they operated "below the threshold of utility" of British weapons.

Since tomorrow's wars are likely to be conducted among the population, technological advantage will only help those who are actually engaged in it. In general, high technology tends to increase effectiveness more in unbroken homogeneous environments (space, air and sea), which is indeed where combat was expected to take place in the past (control of the sea and the sky, with the resultant control of the land environment). But by applying the rule of circumvention, the opponent is now leading us into fragmented, heterogeneous and complex spaces, where modern weapons soon lose their advantages; he is bringing us back down to earth, into the urban environment and among the population, to be exact.

Moreover, another effect of the rule of circumvention –now associated with the new ease of distribution and the easy access to technology– is that technological advantages are only ever temporary, as our societies constantly disseminate their capabilities, which they would otherwise be the only ones to possess. New technology is fated to be first bypassed, then

---

1. On this topic, the Joint Improvised Explosive Device Defeat Organization (JIEDDO) provides a striking example. In view of the constant rise in the number of American victims of IEDs (50% of dead and wounded in Iraq, 30% in Afghanistan: the "number one killer of American troops", in the words of the Chairman of the Joint Chiefs of Staff), this organisation was set up by the American armed forces in October 2003 (in its initial form). It was placed under the command of a senior general and allocated a massive budget: 100 million in 2004, then 1.2 billion in 2005 and 3.4 billion in 2006. However, these mind-boggling sums, invested essentially in R&D, were unable to achieve any real result, with death by IED still the main cause of casualties in Iraq. Faced with this unsatisfactory situation, since January 2007 there have been changes made within the organisation, concentrating on what its director, general Tata, calls its "technology-based focus". Under this new approach, JIEDDO now provides the operationals in the theatre with direct help in analysing the explosive devices and destroying the networks that produce, deploy and detonate them. Given the failure of technology and vast sums of money to resolve this problem, the emphasis is now being placed on "good intelligence": "We need the right intelligence capability. The point is to be able to feel what is going to happen, thanks to intelligence, and then to kill or capture the enemy." (Tata) The comment made by one Senator is quite clear: "JIEDDO and the Pentagon are putting far too much emphasis on technology; it has taken them too long to realise that the important thing is to fight the networks. They have just been trying to solve the problem using technology, but it is much more complicated than that." (Defense News, 3 September 2007, What next for US Joint Anti-IED Efforts?) On 10 October 2007, Defense Secretary Gates confirmed the need for this re-orientation: "One of the challenges facing the Army will be how to incorporate the latest in technology without losing sight of the human and cultural dimensions of the irregular battlefield. For example, we have spent billions on tools and tactics to protect against IEDs. Yet, even now, the best way to defeat these weapons –indeed the only way to defeat them over the long run– is to get tips from locals about the networks and the emplacements or, even better, to convince and empower the Iraqis to prevent the terrorists from emplacing them in the first place."

overtaken: while technological advantages are rarely decisive, they become even more fleeting because a technological surprise only ever works once in the same theatre. If we seriously believe in the possibility of a war between equal, strong parties, then it is time to train to face an opponent who, if he chooses to fight, will have roughly the same capabilities and the same dominance. Napoleon was beaten by his own methods, and France by the massed conscription it had invented. Without any doubt, war will invade the new areas and the new fields created by man. In the short term, future warfare is likely to spill over into space, and then to other new physical and technical areas that have been discovered, but the change to war itself will be no greater than happened with the introduction of air power. Undoubtedly, space will become "militarised", with the deployment of actual weapons, rather than simple support systems for other operations. As a new element of the grammar of war, space will become a new dimension of its environment: shaped firstly by its social and political context, tomorrow's war may also be "war in space", though it will never be "war for space". That war will start and finish on the ground. The character of war will be altered by future technological progress and the face of war will change, but as long as war remains a clash between men and desires the opponent will continue to circumvent and to adapt. Confrontations between people will continue to be resolved where people live, in the heart of populations and societies: "Future high technology cannot become a synonym for the future of warfare."[1]

Indeed, technology seems nowadays to have something of a levelling effect. Not only is technology available more readily (globalisation) and more cheaply, it has become much more user-friendly. In most cases, the latest civilian technology is better and cheaper than the latest military technology. It has become much easier to use, making it possible to benefit from the latest technological wonders with no need for extensive training. This makes it readily accessible to irregular forces. Minor technology transfers can significantly increase the risk and lead forces to adopt protective measures out of all proportion to the actual threat. Since April 2003, it is the use of small quantities of the most advanced technology that has enabled the Iraqis to make life difficult for the American and the British armies. The information revolution –of which we had such high hopes for consolidating power outside the constraints imposed by strategy and the reality of conflict– has actually produced highly negative effects for the most powerful nations of the past: by breaking the traditional link between power and wealth, it is re-distributing power among the irregulars, producing a levelling effect.

---

1. Gray. For more on this subject see his book, Another Bloody Century: Future Warfare

In its fight against the Israeli Army in the summer of 2006, Hezbollah made good use of the latest technology, deploying the tandem-charge RPG-29 rocket or the AT-14 Kornet-E missile, which is known to be capable of penetrating up 1,200 mm of armour at 5,000 m, or even with its direct strike on the corvette Hanit, with a latest generation Silkworm C-802 missile. At the same time, it also bypassed the Israeli high technology. Hezbollah used the population as a form of new technology, turning it into both an offensive and a defensive weapon, the "new armour" of irregular warfare. Exploiting its environment, it is completely based within the population, imitating normal civilian life and systematically causing collateral losses and damage. Dispersion, camouflage and integration into the population make it quite impossible for the Israelis to distinguish between enemy and neutral. Even the most modern sensors are unable to differentiate inoffensive civilian vehicles from those being used by combatants. By using low cunning to get around high technology, Hezbollah made excellent use of the asymmetric situation.

We can be quite certain that the inscrutable enemy of tomorrow's war will easily be able to get round the best that modern technology can offer. There is not a single high-tech device that can compete with the mind of a kamikaze combatant: our intelligence systems cannot locate him, our arsenal offers no deterrent and, all too frequently, our soldiers cannot stop him. Our finest technology and our greatest theories are completely demolished by the strength of the faith of the suicide bomber.

In a more general sense, it is the very nature of future wars that has led to a change in the position of technology. The sort of war that has shaped our thinking and our weapon systems had become primarily a technical dual, removing the connection between the political dimension and the military act and confining it to the overall deterrence operation. This produced the "exponential technological growth", in pursuit of technical perfection. As we now see future wars, there will be a return to political primacy, and a restoration of the connection between the military and the political. The role of exponential technological growth is losing out to a revival of the political aspect. If we observe our soldiers operating in a crisis, it is clear that their effectiveness owes far more to their abilities than to their equipment, in the strictest sense.

THE RELATIVE NATURE OF DIGITIZATION

Remarkable progress in information technology, together with equally extraordinary advances in precision and detection, gave rise, first in the United States, and then in Europe, to great hopes for change in the nature of warfare: it would be revolutionised. The idea that real war would be a network-centric undertaking was the dream of RMA, and later of Transformation, and was anchored in the notion of a strong, unipolar invulnerability, For the modern army, the only thing that counted was Network Centric Warfare (NCW). Their strength multiplier effect and the fact that they enable actions to be conducted in parallel (rather than only in sequence) across the whole battlespace more rapidly than the opponent should have allowed these networked operations to support the myth of the rapid war, in which it is possible to impose one's wishes on the opponent, from a distance, before he has even been able to express his own ideas. Unfortunately, NCW focussed entirely on a single aspect of combat, not on the war as a whole.

The limits of digitization

This "technician's" development, based more on the attraction of new capabilities than on the observation of changes in the nature of conflict, soon reached its limits. There were three main reasons for this. The first relates to the very origin of NCW, which lies in the American managerial view of the world. It was an attempt to transfer business practices to military affairs, based on the underlying concept of the close links between the nature of war and the nature of trade. Admiral K. Cebrowski –one of the first advocates of the idea, who founded and then led the Office of Force Transformation[1] within the Pentagon– believed that: "Nations make war the same way they make wealth [...] Network-centric operations deliver to the U.S. military the same powerful dynamics as they produced in American business."[2] However, the prime purpose of civilian systems is to improve understanding and knowledge of the internal situation: stock and variation, in and outflows, tension, etc. Focusing inwards, these systems attempt to improve performance, and are in no way concerned with opposing production systems or any directly harmful acts by opponents. Thus NCW was initially designed, like the rest of the US armed forces of the 1990s, for a world with no real adversary and no circumvention. NCW requires a certainty and a predictability which, while they may sometimes be found in the world of business, do not exist in war, particularly not in the wars of the future. The shortcomings of this initial error are still felt today.

The second reason is that the initial thinking was carried out rather like a high school maths exercise in which, to simplify the problem, only

---

1. This office was closed in 2006 as a result of the questions arising from the inadequacy of Transformation in the face of new information on probable future engagements.
2. Quoted by Kagan, p. 258.

some of the variables are altered, all other things remaining equal. Two basic rules of life, and thus of war, were ignored: on the one hand, adaptation and circumvention and, on the other, the inevitable cascade of counter-reactions. Network-centric warfare was primarily devised to tackle an opponent unable to respond to the ability of Western forces to see everything and to destroy all they can see. In other words, to deploy highly sophisticated weapon systems against a not very bright opponent – such as one who had stuck with the fairly basic weapons held by the Warsaw Pact to fight an industrial war and who had never opened a copy of Defense News or Jane's. When confronted with the real world, the system performed nowhere near as successfully in theatre as it had in the PowerPoint presentations. At the major arms fairs, it is striking to see how many presentations of new weapon systems still show ever more sophisticated Western systems deployed against stupid columns of T72s emerging from the woods, as if 1989 and the break-up of the Warsaw Pact had never taken place. Unfortunately, neither the Serbians in 1999, nor the Iraqis in 2003, nor Hezbollah in 2006 waited patiently for the so-called "3rd generation" war to be declared against them. In sociology there is a concept known as "creative prediction". It would be better, in terms of both strategy and concepts, to know that what is predicted is unlikely to occur This would be more a sort of "destructive prediction".

The third reason is that, if technology changes anything, it is the forms of war rather than war itself. It changes the "subjective" nature of war, not its "objective" nature, to take Clausewitz's useful distinction. In this sense, the English language offers a useful distinction between "war" and "warfare", though the two are often used interchangeably. Warfare does not change: irrespective of any technological advances, warfare remains a form of social behaviour, a political act consisting of a dialectic relationship between two independent free wills, each seeking to avoid being constrained by the other. Using once again the words of Clausewitz, while the "grammar" of war may change, its "logic" will always remain political.

A more restricted input to tomorrow's war

Warfare does not change: at the heart of any war will always be man – the man fighting, the man we are fighting against, the man for whom we are fighting. Man is an essential and permanent part of the nature of war, no matter what arms and technology are used to increase its inherent force. Looked at from this point of view, there is nothing really new in NCW; digitization is simply one more tool to be added to the long list of those used over the centuries. As early as 2003, the British, whose closeness to the United States allows them to be more critical than us, very rapidly abandoned the idea of NCW, replacing it with the concept of Network Enabled Capabilities (NEC), thereby putting the technological aspect into perspective; war was no longer to be network-centric, rather the role of the network was to facilitate the conduct of operations. Little by little, common sense realised that no-one has ever seen a "kilobit" blow up a bunker! Having initially been overwhelmed by the brilliance of NCW, French doctrine has now fortunately followed the British example, albeit under the ugly, yet more realistic, title of "infovalidation" (*infovalorisation* in French).

The digitization of the battlefield undoubtedly had a multiplier effect on the conventional warfare of the past, but this effect, while it cannot be described as negligible, will be far less pronounced in the conflict environment of future wars. It will certainly transform some types of action –belonging to the inevitable but brief moments or phases of coercion operations– but, in most cases, the modes of action of the forces will simply be optimised. While a war of destruction can be organised in a network-centric way, a war of persuasion or of influence cannot. In the new form of conflict, it is not a matter of winning the battle for time against an opponent who has this advantage on his side; the asymmetric adversary is in it for the long term, thereby rendering the speed of a network-centric approach irrelevant. In the decisive phase of future conflicts, the stabilisation phase, there will be very few "sensor –decision-maker– operator" loops to be closed as fast as possible and few automatic decisions to be taken. Speeding up the OODA loop offers a technical advantage against a similar opponent, or against a dissymmetric opponent for whom it restricts the damage he can cause. In general, however, it makes little sense, even tactically speaking, since the greater part of the fight against an asymmetric opponent will be outside the field of NCW and, in attempting to control violence, discrimination and reversibility are more important than response time, and destruction may even be counter-productive.

Making the best possible use of digitization

War is, of course, too complex a phenomenon, with too many dimensions, for information superiority alone to provide a magic solution. War involves much more than simply aiming at targets, even if this is the idea behind NCW, designed for use against an inert adversary, with the Transformed forces allowed to make the best use of their wonderful new systems.

However, even if no level of information superiority can compensate for weakness of morale, lack of discipline or inadequate training, refusing to join the resolute march towards digitization is not an option. Whether we like it or not, the train has left the station. Staying behind on the platform would mean being part of a second-rate force, the back-up troops destined for the lowliest tasks –the window cleaners– compared with the major Western forces, with whom we would no longer be compatible. We need to advance resolutely –without losing sight of the real input and the potential negative effects– in directions which fit in with our own vision of warfare. It will therefore be necessary to adapt to new circumstances a form of digitization designed for very different contexts and purposes, when the aim was to increase accuracy against a tangible opponent, and when the presumed omnipotence of the air force (after the First Gulf War) made the ground resemble a series of targets, much as a dartboard must appear to a darts player.

Initially digitization was a targeting tool, intended also to speed up the OODA loop; that has all changed. Its effectiveness now has to be seen in the new contexts of military action, with their different aims and increased complexity. In June 2007, the Chief of the French Army Staff gave a clear indication of the objectives that can reasonably be set in present circumstances: "The digitization of the battlespace should no longer be designed primarily to enhance the decision-making capabilities of central echelons, but rather to provide the tactical echelons with all the information and fire power required to grasp any opportunities that arise... digitization should be seen as a system which strengthens the capabilities of the echelons by improving the analysis and transmission of the information and making it easier to provide the support needed."

An undeniable contribution

What we do know, is that digitization definitely offers added tactical value, in that it contributes directly to economy of forces and to adapting to circumstances, the keys to success in war.

Regarding the economy of forces, it is true in the way the term is used by Foch: an improved knowledge of the situations and the positions allows a reduction in reserves, while concentrating efforts makes it possible to do more, at least initially, without the need to regroup assets. However, these gains are limited by the need to avoid too much reduction or dispersal of the elementary units. From both a psychological and an equipment point of view, excessive reduction has a weakening effect; there is thus an optimum level of dispersion, below which one should not slip, even when technology makes further dispersion possible. The modules have to be composed in such a way as to make them "tactically unsinkable" as they weather the inevitable storms and squalls found in any theatre of operations. Furthermore, digitization has no effect on human psychology, which is still the ultimate judge of the new theoretical capabilities. No matter what technology is deployed, men will remain men, and fear will always be a significant component of human behaviour.

The new "dilution" of resources made theoretically possible by digitization, thus also runs into problems with the human need to stick close together in difficulty or danger.

Economy in terms of the volume of forces in a given theatre is also possible, as digitization improves efficiency, in other words producing a better operational result for the forces deployed. By making it possible to catch up, at least partially, such increased efficiency is particularly welcome at a time when our reduced forces seem generally under strength given the new tasks they have to perform in modern conflicts and the increasingly vast areas they need to control.

With respect to adaptation, once the initial heavy problems linked to the installation and organisation of the systems have been overcome, digitization should allow greater flexibility. In the past, the problems of autonomous navigation meant that units were subject to major restrictions in terms of their formation and that, having engaged on a manoeuvre, reversal was impossible. In the future, the manoeuvre will be far more flexible; dispositions and structures will be adaptable, as the organisations will follow the reactive nature of the networks, rather than the constraints imposed by pyramidal links. This will make it much easier to adapt to circumstances and seize opportunities by combining fluid sequences of events, the sustained pace of engagements and the spontaneity of actions.

The challenge of confronting reality

Nevertheless, the move from experiment to application calls for caution. Besides the psychological problems which are bound to arise from the loss of habitual references and the frequent change of subordination, two other problems remain unresolved. The first arises out of the artificial merging on the computer screen of an accurate and up-to-date image (the "friends") and an image that is always out-of-date and inaccurate (the "foe"); at a tactical level, this risk can be considerable, as the American ground troops found on several occasions as they fought their way to Baghdad in April 2003. The second critical problem, leaving aside the difficulty of squeezing vast amounts of information down narrow "pipes", relates to the environments in which modern combat takes place. The location of the war has changed and the open spaces for which digitization was designed have been replaced by fragmented and restricted settings, notably urban areas. The opponent draws us into these places, knowing that they wipe out our advantage; our systems lose much of their strength.

The risks allied to digitization are far from negligible. They include, in no particular order, information overload, with the lower echelons drowning in irrelevant information; interference by superiors tempted to conduct other people's business rather than their own; the "virtualization" of war, cited as one of the causes of the difficulties encountered by the Israelis in July/August 2006; blind faith in a reference image issued by a superior level, but inevitably inaccurate at the lowest tactical echelons; the imposition of too high a pace, leading to rapid weakening as a result of the disconnect between staffs able to speed up operations at will and troops operating in trying conditions; excessive initiative on the part of subordinates, increasing the need for a command style based on initiative justified by intent. On the subject of pace, a paradox has rather burst the bubble: speeding up the manoeuvre has made it difficult to supply the forces with the information they need. General James Mattis, the then commander of the 1st Division USMC, remarked that, although the digitization of the systems had been completed, during their entire attack on Baghdad, his division received no worthwhile information from the superior staffs, nor from the national intelligence echelons.

The risks include cumbersome procedures, leading to a reluctance to use the tools and, more importantly, the inescapable problem of interoperability, not only between the various systems used by the different armies attempting to cooperate, but also within the individual armies. Technological developments make it necessary to change equipment, but the cost of replacement means perfect homogeneity will remain a dream. Like Sisyphus and his rock, as soon as an army introduces digitization, it is condemned to seek to overcome the "digital divide" – an impossible task. The trap, into which some of our allies have fallen, is also to confuse experiment with reality. The trials are usually carried out in a totally virtual environment; however, when they transfer to the real world, the scenarios used generally involve an essentially passive opponent, incapable of destroying or bypassing our systems, which may not always be the case.

Technology is simply technology; it cannot change war, which is primarily dominated by human factors. Concentrating too much on the problems of "information plumbing" is not necessarily the best way to revitalise a force for its role in future wars. Tying a whole defence system to dazzling innovations in one area, and the tactics they allow, can lead to severe disappointment. At the end of the 1950s, the Americans adopted the single strategy of "massive reprisals", in the belief that all other forms of combat had disappeared. This led to a weakening of the armed forces and caused major difficulties in Vietnam. There is a similar risk with the concept of Transformation and the myth of Information Dominance. It would be wise for the impact of successive waves of new technology to be put into perspective. What is important, is to advance at a reasonable pace, in the understanding that digitization is a means, not an end, and to reject the invasion of the virtual for its own sake. We must guard against the vision of future combat being against an opponent who has not been considered fully in our thinking. General Rupert Smith spoke of "the clear danger of knowing more and more about ourselves and proportionally less and less about the enemy". The purpose of digitization is to improve operational capabilities, not for their own sake, but to enable the land forces to contribute to achieving the desired political effect.

"Cyber vulnerability"

It is highly likely that digitization will give a boost to the technical effectiveness of our forces. Simultaneously –and inevitably, since every strength is accompanied by a weakness that can be exploited by the adversary– it burdens them with a new form of vulnerability, one which will increase over time if we do not take care. In the absence of an actual confrontation with a similar opponent, our natural inclination is to place our trust in our new tools. There is, therefore, a risk of losing the knowledge and the equipment needed to operate in "fail safe" mode, that is to say, if our systems, for malicious or other reasons, are no longer able to function correctly. It is essential to be fully aware of the new cyber vulnerability of our forces.

In our major exercises, our digitized forces fight opponents who are incapable of destroying our networks or attacking our servers. This will not always be the case. The enemy we are lacking in our training camps is the real enemy, whose chief wish is to disrupt our organisations, our systems and our plans. If we are serious, we must recognise that in conducting a high-intensity digitised war against a peer competitor, such an opponent will only accept this form of combat if he also has comparable capabilities or if, having taken appropriate countermeasures, he is confident of being able to wipe out our superiority immediately. If we attack a third-rate opponent, the accuracy of our weapons will force him into asymmetry; other, much less sophisticated, means will be required to compel him to submit to our will. In both cases, this future war is liable very rapidly to become a primitive war, in which success will depend more on moral forces, human skills and the accumulation of localised victories than on a top-down approach, a major systemic battle and satellite networks.

Cyber vulnerability is a concept with a great future. In the spring of 2001, the international community became fully aware of the fragile nature of its information systems, when Estonia suffered consolidated attacks on its web servers. However, according to the experts, this attack was simply one of thousands that had gone unnoticed. As soon as the attack –which may have been linked to the tension in relations between Estonia and Russia– began, the websites of the Parliament and the Ministries of Finance and Agriculture immediately became inaccessible. So too did those of the two main Estonian banks, which had dramatic repercussions since over 90% of local banking services are only available on-line. One might be tempted to think that Estonia is less well protected than other countries. But what about the Chinese "cyber-pirates" who apparently managed to freeze the computers of the US Defense Secretary in his office in the Pentagon? Or similar attacks on Germany, which recently led the Chancellor's office to complain to the Chinese prime minister that spy software had been installed in its offices in Berlin?[1] General James Cartwright, Head of US Strategic Command, warned recently that: "The confusion and disorganisation that would be caused by a cyber-attack on the US nerve centres would have a psychological impact comparable to that of a weapon of mass destruction."[2] Our major systems –and everything that depends on them– are more and more vulnerable; it will probably be necessary to move towards smaller and more compartmentalised systems, less interdependent and thus less vulnerable to the systemic attacks that our opponent –learning from our own principles– will undoubtedly launch.

In terms of information systems, the technological progress has also become something of an Achilles' heel; like the others, these systems are destined to be either attacked or bypassed. The other side, however, has no hesitation about announcing its intentions in all areas. Beijing has clearly demonstrated that it can jam or destroy our satellites from the ground. The comparative technological advantages, on which we base our conviction that we will win in the future, could well be very short-lived in the event of a major war.

It is by looking at his feet that the true power of a colossus can be judged.

A NEW PROCESS FOR DESIGNING WEAPON SYSTEMS

These reflections on the sort of war we are likely to be called upon to fight, and the realisation of the urgent need for equipment that can constantly be adapted, lead us to re-examine our design and specification processes; these are still tied to a way of thinking that became obsolete eighteen

---

1. Financial Times and *Der Spiegel*, quoted in *Le Figaro*, 5 September 2007, p. 2.
2. Idem. General William Lord, first head of Cyber Command, created recently within the US Air Force to develop, prepare and conduct the defensive and offensive cyber-war, believes that: "We need to think about *weapons of mass disruption*, not just weapons of mass destruction." *Le Figaro*, 29.10.2007.

years ago. At present, the contradiction between the extreme volatility of events and the cumbersome nature, or even inertia, of the processes involved in designing and producing weapon systems seems to have brought us to a sort of impasse. We need to change our thinking. As General Rupert Smith observed, the opponent soon learns not to present the sort of targets suited to the design of our weapons and the way we operate them. Experience and history also teach us that, generally speaking, arms are almost never used to produce the effect for which they were designed: in reality, they seldom satisfy the needs. It is almost always necessary to adapt either the equipment or the way it is used. This calls into question the principles of highly detailed specifications and a very long gestation period.

With the exponential increase in the speed of technological development, progress in the technical performance of weapons is also steadily speeding up; nowadays it does not take fifteen years for a piece of equipment to become totally out of date. By the time it is taken into service, even our most modern equipment seems no longer to satisfy the requirements. The first Rafale flew in 1987, but it will have taken over twenty years before it makes its first operational flight in Afghanistan; by that time, the original design will have undergone many significant changes in order to adapt to changes in engagements. The Leclerc tank, designed right in the middle of the Cold War, is currently being modified; the new PVP armoured liaison vehicle and the VBCI infantry fighting vehicle also have to be modified before they can even enter service. This situation seems to call for equipment which is perhaps somewhat less sophisticated, but which can be designed and produced rapidly and replaced more frequently, or equipment with a system of successive incremental improvements incorporated into the design. The best form of adaptation is one built in from the outset, as any system of adaptation not incorporated into the design phase of the programme is likely to be more costly and less successful. Astonishingly, under the influence of NATO –that machine insisting on producing to meet standards–, the enormous American military budgets and the uncertain race for interoperability, Western armies have always opted for the opposite approach.

Other concepts may also contribute to facilitating the adaptation of equipment to circumstances, while, at the same time, rationalising budgetary choices. For some specific missions, the best solution can be a more or less disposable, "off the shelf" product. Some countries, such as the United States, the United Kingdom and Germany, use this method, which combines financial and operational interests. The same applies to the continuous loops linking designer, user and producer, which is a direction that could well be adopted by current processes, which are both too long and too linear in their approach.

This raises the question of the true benefit of very long-term thinking about forces and their equipment. Of course monitoring technology, basic research, building up potential, following trends and open reflection are absolutely essential: but why is designing a weapon system that we intend to be able to deploy only in 25 to 30 years? First of all, this will ensure that, as a result of the law of action and reaction, our future opponent will have in place the weapons and counter-measures to render them of no value. What use would a weapon designed in 1900 have been in 1925? What use would a weapon designed in 1925 have been in 1950? What use would a weapon designed in 1950 have been in 1975? Here this reasoning must stop, as it brings us to the Cold War, a period which saw the enemy cease to evolve, extended the life of weapons and, sadly, still all too frequently underlies our thinking today. However, it is feasible to ask: what use will a weapon designed today be in 2025? The answer is surely: very little, as the speed of both change and obsolescence has increased considerably.

The root cause of the problem is that our design/production process is frequently far too long: we are striving to find a perfect, stabilised and universal solution, which just does not exist. Our opponent, on the other hand, uses models, seizes the latest technologies and is prepared to accept an interim solution, which he can then gradually improve. One could say that, while our operational cycle is faster than that of our adversary, his technological cycle is much more rapid. Thus technology is both one of our weaknesses and one of our strengths; in an intervention the opponent will be able to bypass our best systems and exploit our vulnerable points before any confrontation takes place. It is perhaps time to re-discover the "Logan spirit", aiming to provide the necessary, and only the strictly necessary, at a controlled and restricted cost, allowing 80% of the requirement to be met rapidly and at a reasonable cost, and to abandon the perfectionist approach, which satisfies 120% of the requirement after 15 years, at a very high cost for the marginal additional capacity.[1]

This has now, at last, become the official line of the Ministry of Defence. Speaking on 11 September 2007, at the Ministry of Defence summer school in Toulouse, the minister, Hervé Morin, questioned the "justi-

---

1. Speaking on this subject on 23 May 2007, at the Defence Research laboratory of *IFRI* [French Institute for International Relations], Mr Marwan Lahoud, the then CEO of MBDA, said that, from the operational point of view, it was a question of accepting a tighter dialogue in order to arrive at the correct expression of the requirement, and thus the minimum acceptable level for the military specifications. In his opinion, this means accepting into service weapon systems that meet only 80% of the desired requirements, but enough to accomplish the task allotted to them. Once the system had been tried out in the real circumstances, and if necessary, the remaining 20% of the specifications could be introduced as part of an upgrade. Obviously, this possibly would need to be incorporated at the design stage.

fication of the perpetual technological quest", and expressed the desire that "when making their decisions, politicians could be offered the alternatives of a highly sophisticated system including all the latest technology and a system which, although less sophisticated, nevertheless meets the real requirements imposed by the threat". He pointed out that the "leap of hypertechnology –the 5 or 10% that represents the ultimate technological effort in any programme– can increase the cost of the programme by 20 to 30%." The solution proposed by Bruno Rambaud[1] should provide a solution to this problem; he believes that the cost of technology should be measured in terms of the additional operational value gained, in other words, there should be a full analysis of the value.

Of course, present day Western technological superiority is the result of a policy of investment in R&D over the past decades. It is essential to continue in this vein, as the R&D policy of a nation at any given time partly determines its international status twenty or thirty years later. It is therefore crucial to support innovation and fundamental research. Nevertheless, by limiting armed forces to reduced formats, insufficient for the new strategic environment, the marginal cost of the excessive performance requirements affects their overall effectiveness.

These requirements are not contradictory, provided we look at things differently. We must maintain basic advances while, when it comes to equipment, rejecting perfection and preferring the realistic to the ideal.

ADAPTABILITY: THE NEW CONDITION OF UTILITY

Survivability demands adaptability: ask the dinosaurs! This is true of life in general, but particularly true of life at war, since being at war means attempting to curtail the survival of the other side. This is an eternal truth, but one that has been rather forgotten as the quasi-predictability of the conditions of engagement of the Cold War dulled the importance of adaptability. This can only be based on the sense of reality: on observing it, accepting it and taking it into account.

It is well worth taking the time to read what has been said on the subject of adaptability by General Rupert Smith. Looking back over his operational career and the several senior commands he has held, he reaches an important conclusion: every time he deployed on a new mission, he was only able to make the force he was commanding useful once he had reor-

---

1. Ideas reported in *Le Figaro*, 26 September 2007. Bruno Rambaud is the Senior Vice President of Thales Land & Joint Systems, and President of *GICAT (Groupement des industries françaises de défense terrestre)* [French Land Defence Manufacturers Association].

ganised and modified it, in other words, adapted it. The need to adapt was driven by the –ever changing– nature of the opponent and his circumvention actions, the constantly varying environment and the surprises that are an inherent part of working in a coalition. In his opinion, our problems are not caused by the fact that we are preparing for yesterday's war, but rather that we are preparing for the wrong war.

There are really only two things of which we can be certain with regard to the wars of the future: on the one hand, the unpredictability of the circumstances under which they will start and, on the other hand, the fact that they will be different from today's conflicts. There will be "strategic surprises", which really will surprise us. This risk does not mean that we should hang on to the models and arsenals from the past, because we also know that this surprise will be a surprise with no restrictions, and that both the war and the conditions will surprise us. Thus, what counts today is our ability to respond. More than ever before, it is not the ability to plan and conceive which is key, but rather the ability to adapt; firstly because it is quite impossible to estimate accurately the occurrence of future types of engagement, secondly because it will generally be necessary to fight a reactive war and, lastly because, in the course of such a reactive war, all command levels will constantly be called upon to respond, with very little time for preparation. In the future, it will be increasingly necessary to know how to adapt our equipment, which will have to have such adaptability built in at the design stage. This is a matter not only of culture, structures and procedures, but also of intellectual, structural and financial "reserves". Today, it is primarily a matter of vision and intention.

There is also an important question of flexibility within operational organisations and hierarchies, and here digitization will be of great benefit. In the future, success will be not so much a question of technical capability, but rather of the ability to ensure the constant adaptation which must form the basis of the style of operational command and troop training. To achieve this vital level of adaptability, and to allow clear decisions to be taken calmly in the face of operational events, will require a strong faith in man, the consolidation of the common culture and the construction of flexible systems.

We should be in no doubt that tomorrow's war is most certainly war, and that war is always a "chameleon", to use Clausewitz's excellent image. This means that the war fought will never be the war predicted and that it will adapt to suit the circumstances. It also means that it is better to focus on circumstances than on tools or theoretical models, which are only valid if they can adapt to circumstances.

Success in future wars will be less a question of equipment –it will be bypassed, whatever happens– and more of ensuring that we have a constant ability to adapt.

CHAPTER V

TWO SPECIFIC CONDITIONS

THE FUNDAMENTAL ROLE OF PROTECTION

In the new operating conditions, protection has become a fundamental aspect of effectiveness. Somewhat paradoxically, when war was seen only in terms of a climactic confrontation between two blocs, force protection was a much less sensitive issue, for two reasons. Firstly, our vision of war was of an absolute war over vital interests which, under the strong influence of Clausewitzian thinking, we believed justified and inevitably involved massive losses. Secondly, the lack of adequate protection only changed the least important dimension of military action, its operational effectiveness.

Circumstances have now undergone a dramatic change, reflected in the renewed importance accorded to protection. As well as affirming its influence on the technical effectiveness of weapons, protection is also a condition of the freedom of action of governments, who must remain capable of acting in a crisis today, but also tomorrow. Just as future wars have rediscovered a political substance, which had faded into the background, force protection has now taken on a fundamentally political dimension. Protection has a new status, somewhere between "means" and "end"; it reveals a new complexity since, having cast off the passive aspect which marked it for so long, it now appears as the convergence of both active and passive and direct and indirect measures.

Force protection has become a political issue because it is the very condition of political freedom of action. This is well illustrated by the withdrawal of the French and American contingents following the bloody attacks they suffered in Beirut in 1983 or, ten years on, by the withdrawal of American troops from Somalia after the events later portrayed in "Black Hawk Down". Even in engagements which may not appear to be critical, political freedom of action –now, but also in the future– has become a key element of force protection. Furthermore, since we are no longer facing the possibility of a single, massive engagement –the all or nothing of a game of poker– it is essential to preserve the forces for use in the series of crises which will undoubtedly take place. Future wars will be fought by forces which are not "disposable". Plans for the use of the force must take account of the absolute need for its preservation –and thus for its protection– because the same scarce and expensive forces, and the same equipments drawn from a non-renewable supply will have to operate in a succession of theatres of engagement. This is especially true at present, since forward defence results in a string of lengthy and overlapping engagements; Western forces cannot afford any significant losses. These requirements have a considerable effect on the way we conduct a war; instead of using the force to achieve an aim at any cost –which was the philosophy during the absolute conflict of the Cold War– we use methods allowing us to safeguard the force.

At a tactical level, in the current crises, protection remains an issue linked to effectiveness; it is a question of operating in the area, among the population, with a view to establishing contact and de-escalating the level of violence. For reasons of credibility and confidence, it is essential for the force to be able to protect itself; If it is unable to do so, the population would have no reason to believe that, if they were to side with the force, they would be protected from the other party. Such a loss of face would harm the credibility of the force, but also of the project it represents. In other words, whether it is "attacking" or "being attacked", the force must "win".

There is, clearly, a contradiction between the protection of the force and its ability to fulfil its contact mission. If it has suitable protection, the force can intervene without needing to respond to violence by increasing it. However, at the same time, the physical protection measures risk isolating the soldier from the population, who may then see him more as the "other side". Reducing the footprint, the visual presence which can quickly be seen as oppressive, reduces both the risk of and need for protection, but in order to make a real change to the local situations it is important to be present in significant numbers. A badly protected force hides away, cuts itself off and does not understand the situation; it easily gets caught up in a spiral of ineffectiveness or violence, terms which are frequently confused. There again, the effect of losses is felt immediately; this is demonstrated by the changes introduced by UNIFIL following the attack which killed six Spanish soldiers in June 2007. We see here the deeply paradoxical nature of protection: tactical effectiveness demands protection, but excessive protection weakens effectiveness. The challenge is to achieve protection through a number of measures, while avoiding compromising the effectiveness of the force, since this effectiveness is itself a part of protection.

Thus, since protection is a fundamental element of the overall effectiveness of the force, this is an extremely complex problem. It has moved from being a simple matter of armour to encompassing all aspects of the life of the force. A soldier on patrol needs to be protected, but so do the logistic support convoys and holding areas, which are increasingly threatened by RAM (rocket, ammunition, mortar) attacks or even chemical attacks, with toxic clouds spreading across the ground, as happens in Iraq.

Such a paradox! Such complexity! This makes simple solutions impossible and turns protection into a consequence, an effect produced by the convergence of a variety of actions and by an overall concept. Passive protection alone is worthless; as long as armour can be defeated, it will always be possible to circumvent passive protection[1] (as clearly demonstrated by the development of IEDs). Even the best technological ideas have negative side-effects, such as the effects of reactive armour on accompanying infantry or, even more so, the effect of jammers on other equipment. However, the limits of the passive approach should not deter us from adapting the vehicles we deploy in theatre, creating mini-fleets of vehicles suited to the specific threats of an engagement.

The limits of passive protection make one thing clear: priority must be given to the active approach, which involves reducing the threat, and thereby altering the overall manoeuvre and the general tactic used. Protection, like security in general, is an overall consequence, with various components (men, convoys, bases, etc.), facing various threats (IEDs, third dimension, rudimentary chemical weapons, etc.) and deploying various families of responses (active and passive, direct and indirect, preventive and proactive) in various fields (doctrine, equipment, instruction, training, etc.) and using various technical tools (general intelligence, early detection, immediate detection, robotics, etc.). Protection requires adaptation, preferably at the same pace as the counter-adaptations being carried out by the other side. Adapting in this way, which is necessarily reactive, must be in line with the fundamental question of the degree of

---

1. It nevertheless remains essential to continue to improve armour. For some time to come, such improvement will consist primarily of making it thicker, and consequently making vehicles heavier. At the end of 2007, the general opinion in Britain was that the average weight of vehicles would increase. The dream of multi-purpose vehicles in the 20 tonne range has been shattered. We now know that for a combat platform to survive on the real battlefield, it will probably have to weigh at least 30 tonnes. It is no longer possible to achieve a trade-off between protection against rapid strategic deployment, which we have also come to understand applies, for a number of reasons, only to a very small section of the force to be projected.

risk to which we are prepared to expose our troops, and for how long. Here the political and military hierarchy are dealing with responsibility and results, rather than resources.

Our experience of the last fifteen years of engagements has also taught us that number is in itself a factor of protection. It allows an understanding of the environments and how to control them; it has a direct effect on the attitudes of both our forces and their opponents. Even though we are trying to reduce our footprint on the ground and –by making use of air transport– that of land convoys, deploying a force that is too small –which has been "slimmed down" too far– puts the force in danger, whatever passive methods it may have available. The scarcity of troops on the ground means that they either have to shut themselves away –thereby making them useless– or take more risks. It also implies a greater reliance on air strikes, which increases the chances of collateral damage and, at a later stage, repercussions on the ground troops. From a tactical point of view, the best way of protecting troops is to have enough. In addition, history shows that poorly trained troops are the ones who suffer the greatest losses; training means time and time means numbers. Thus, while number does not imply quality –which is the result of equipment and training– it is nevertheless an important quality.

An army which is too small to carry out the missions it is called upon to perform is an army which is simply putting its men in danger. We are faced with a moral obligation.

MULTINATIONALITY AND INTEROPERABILITY

While future wars are unlikely to be fought against a state, they will undoubtedly take place in an international environment, in which national freedom of action will be restricted. The multinational nature of all operations in which French forces are involved is already evident; it causes a range of problems. As the Chief of the French Defence Staff said clearly: "Multinationality weakens the deployed force and obliges us to maintain on a permanent basis a level of interoperability with our allies which is both costly in terms of equipment and demanding in terms of procedures. [...] We need this multinationality in order to legitimise our actions but, from a practical point of view, it causes enormous problems. [...] The rules of engagement enforced by each nation create barriers to the cohesion of the force. [...] In effect, even if the effort put in to working together is of value in itself, multinationality is more of a hindrance to conducting operations than an advantage."[1]

---

1. General Henri Bentégeat, *École Navale*, 25 Januray 2006.

It is very difficult for a multinational land force, in which human aspects naturally play a greater role than technology, to be cohesive. Away from the fine words, for some time to come, multinational forces will continue to be a collection of national elements deployed, in most cases, for national political purposes, pursuing their own national aims and interests, with different rules of engagement, and thus operating at the lowest common denominator. Experience shows that, as soon as the stress and intensity of the operation increase, it is tricky to conduct operations which go beyond the level of the largest national contingent.

What is less clear to many people is that the question of interoperability goes far beyond that of multinational units, since it also affects combined arms and joint operation and, given the context of recent engagements, even enters the inter-ministerial domain.

What is also not clearly understood is that interoperability is by no means solely a technical issue: if we truly had the political will to engage in coalition with a partner, we would always be able to find the necessary technical interfaces. There are numerous examples in history, even though we usually prefer to anticipate than to be precipitate. Let there be no doubt, interoperability is first and foremost a question of mutual understanding, that is to say, a question of cultural similarity and compatibility of doctrine. This is where we should concentrate our efforts.

Regrettably, interoperability does not have time on its side: as an all-encompassing concept, with many dimensions and many constantly changing variables, interoperability is more difficult now than it was in the past. Technical problems play a major role here; for example, rapid developments in information technology mean that there are already serious problems of interoperability between various components of the same army. Technological interoperability is increasing in complexity at the same rate as the equipment involved.

However, the prime causes of the difficulty of interoperability are political, and political interoperability is far more important than technical interoperability.

We should not forget that it was the existence of a common project which resulted in the creation of that great standardisation machine, NATO. As it is still understood today, the concept of interoperability developed in both American and European thinking as a result of the similarity of the major strategic projects of the Cold War. Reflecting the need to coordinate efforts towards a common purpose, interoperability was based on the compatibility of equipment and the harmonisation of operational procedures, with a view to improving technical and cultural understanding. This interoperability was translated into NATO standardisation of equipment and doctrine of employment; the similarity of new technologies made this a viable proposition. But this is no longer the case. There is no longer any common project as compelling as the societal project which inspired the interoperability during the Cold War. On the contrary, the instability of circumstances and the notion of coalition make it less essential and reduce the value of any effort, if we do not know in advance under what conditions or with whom we will be called upon to act.

This requires a complete re-examination of interoperability, in order to understand that it is primarily a political goal, and only later a technical objective.

We cannot be everywhere all the time, therefore we must determine where we should concentrate our efforts. Having realised the considerable restrictions imposed by multinationality (as well as its inevitability), we need to identify exactly what is expected in political terms: With whom am I most likely to be operating regularly? With whom do I need to be able to engage what size of force for what type of mission? What is the minimum size of force that needs to remain national? (This for reasons of both technical effectiveness and political visibility, since participation is of no value if by spreading the forces too thinly, they are no longer visible.) So, in summary, with which forces, and at what level do I want to have interoperability?

It is worth noting, in passing, that changes in the international context mean that there is now a mutual need for interoperability. Nowadays, legitimacy implies effective multinationality in deployments and engagements. While the "small" nations need the "big" ones, the reverse is now also true. This gives a "small" nation a stronger role in negotiating interoperability and obliges the "big" nation to take account of the constraints of smaller partners in its own developments, particularly technological developments.

If sought purely for its own sake, interoperability could eventually affect national ability to adapt to the expected conflict environments. There is thus no point in pursuing it simply on principle; it is not a cardinal virtue. It is essential for interoperability to have a purpose: firstly, an operational purpose but, more importantly, also a political purpose. The only form of interoperability that makes sense is differentiated interoperability. Its driving force is the shared nature of the project. Europe could become the best example.

GENERAL CONCLUSIONS

War has not changed. It is still, at heart, a political act comprising a struggle between two independent free wills.

Nevertheless, we are only just getting over the idea, which has dominated for over forty years, that war is a technical confrontation between two arsenals and that the more powerful the arsenal, the greater the chance of victory. The crises we are encountering today are making us aware –but only very slowly– that an accumulation of technical strength may prove to be only an accumulation of political impotence, if we do not take account of the changes in the conditions of employment of our armed forces. However, the fact that even our most advanced weapons are unable to play a significant role in achieving our political objectives should give us cause to reflect and to make it clear –to the military as well as to the politicians who, ultimately, decide on the structure of the force and how it is to be used– that we need to adapt to the new conditions of tomorrow's war, if we wish the armed forces to continue to be of use to the state.

The budget for the French armed forces suffered considerably from the idea of the "peace dividend", which we now know raised false hopes. During the 1990s, the size of the French military capacity was considerably reduced, while the crises and the need to engage were multiplying. The military instrument has been subject to severe pressure, from outside and from within. The crises, however, have not evolved in the same way; we must be aware of what appears to be a growing and dangerous divergence.

A FORWARD-LOOKING ARMY FOR A FORWARD DEFENCE

In the past, in the days of a conventional opponent, a defence could be constructed on the borders, where it would wait for the enemy to materialise. Under the control of states, violence was contained within their boundaries; today it resembles a cancer which will spread if it is not caught in time, with the healthy cells being under threat from the outset, no matter how far they are from the malignant cells. The new forms of violence, in combination with the effects of globalisation and the porosity of borders, make such a "waiting game" a dangerous course, and require, in contrast, the active construction of a stable environment outside national territory.

Strategists warn us of the pressing need to defend "in the deep area"; if not we risk being overwhelmed. We thus need to defend and stabilise "at the front", on the "external circles", while being prepared if necessary to defend while withdrawing to the "intermediate circles", before putting up a "firm defence" of the innermost circles. The aim of defence is to ensure that security is established first "on the front line". The first line of pro-action often lies far beyond national boundaries, close to the "black holes" that must be first contained, then reabsorbed. Here prevention, in its many forms, plays a key role. Early intervention, outside national borders, is required to eliminate the origins of violence, reduce the tension and instability that leads to crises, and take control of the process of conventional or nuclear proliferation: no modern legal or security Maginot Line could possibly offer protection for any length of time against external violence, and its modern manifestations, such as terrorism or organised crime. Both realism and idealism oblige us to take action; absolute security in island-Europe is not possible in a world in crisis and without security.

Already the continuity between security and defence means that we can no longer afford to "wait and see".

In the light of spreading collective violence, giving in to the temptation to dig ourselves in would be to shut our eyes to danger. If we do not make a move to tackle this violence, it will surely come to our doorstep. We have no choice other than to take a resolute stand in our turbulent world and to accept that we have a permanent commitment to long and painful operations. The values of France need to be defended in the Hindu Kush and on the banks of the River Zaire.[1]

There is no doubt that France needs a "forward-looking" army for a "forward defence".

If defence is not pro-active, it is likely to fail: protection –an essential strategic function, which simultaneously represents both the first and the last aim of the defence apparatus– implies prevention. This second strategic function consisting of a range of components, including nuclear deterrence, conventional deterrence (which needs, in fact, to be strengthened, as its credibility has taken a serious knock in current operations) and pre-

---

1. In the words of a German Minister of Defence.

positioning. Nevertheless, intensifying defence only on national territory could result in threatening that which it purports to be defending, by gradually undermining common freedoms, as demonstrated by the ambiguous nature of certain anti-terrorist measures introduced elsewhere. If France and Europe decide to take charge of their own security requirements, this will involve a long-term development of the third, and last, strategic function, which is the stabilisation of those areas likely to export violence. They will regularly be called upon to deploy forces outside national territory, either independently or in coalition, in order to resolve intra-state crises and assist in improving the general situation. In other words, according to one of the key strategic principles, one should defend as far forward as possible and always "in the deep area", on the "external circles", while being prepared, if necessary, to defend while withdrawing to the "intermediate circles" before putting up a "firm defence", if required, of the last circle, that of the internal protection of the territory.

In this context, a solid capability for action on the ground will reinforce the French position in European decision-making. Rather than owning a range of symbolic tools, unsuited to a large part of the modern conflict spectrum, having the right tools for military action, for promoting peace in society and preventing terrorism, will be far more relevant and will count for more, as it will make it possible to tackle the real issues.

In tomorrow's war, the army will be the force of the nation, and man the force behind the army.

ADAPTING FORCE MODELS

Even if it would be irresponsible to abandon the assets needed to respond to the re-emergence of a major military threat, if we were unable to prevent this happening, over the coming twenty-five years or so, there will probably not be any conventional armies to rival the European or Atlantic coalitions.[1] We must therefore retain our conventional capabilities. Not only do they play an essential role in preventing the return of this type of threat, by deterring the opponent from engaging in a race for power, they also strengthen the indispensable coercive diplomacy, by offering a range of threats in support of the deterrence, and enable us, if required, to use coercive force in our external operations.

---

1. Even if such armies were to exist, they would not attack along classic principles, "strong versus strong". Bypassing our comparative advantages, set out in defence publications for all to read, and avoiding our strength, they would attack our vulnerable areas and exploit the areas we consider "off limits".

However, this excess of strength has negative side-effects, which generate both a rejection of the societal models that produced it and the improbability of being involved in the type of conflict to which it is best suited. By reducing the likelihood of large-scale, anti-force actions, in the lack of balance between the arsenals it, paradoxically, comes up against the very limits of its utility. By dissuading, it discourages. The opponent, always keen to avoid conventional violence, seeks to find a means to achieve his political objectives in the new spaces in which confrontation can take place. Having moved from being capability-based to being objective-based, success in war is no longer tied to traditional force ratios. Resolving a crisis requires not only powerful military tools, but also political, diplomatic and social tools, to be deployed across all these registers by forces able to operate in future wars.

In fact, the new contexts alter strategic activity, and even out the advantages gained from advanced technology. Strength is replaced by influence. It is now less a matter of conquering space, and more of placating hearts and persuading the population to support the project being proposed to them. It is no longer the ability to destroy that is key, but rather the ability to achieve political control of the space and, through the carefully regulated use of force in an action considered to be legitimate, to create the conditions needed to establish a new social contract. It is also necessary to display great determination to resolve the crisis; engaging land forces, and thus endangering the lives of the soldiers of the intervening nation, is perceived very differently from the disembodied actions of stand-off weapons.

Since 1945, most conflicts have taken place within states, and this trend continues to grow. This calls into question the concepts and models designed for inter-state wars, and reduces the contribution made by high technology to the new, diplomatic approaches. In the new conflict landscape, the quantitative, rational and classic certainties of the 20th century are gradually losing their validity: the deregulation of war has given rise to forms of crisis in which man is once again at the heart of our defence systems. Armies can no longer create solely homothetic models. Priority must be given to the principle of adaptation, by introducing the internal and external judgment necessary to construct forces capable of participating in coercion operations and also of delivering the required political result in the field. As a result of the interdependence of states, the permeable nature of societies and the globalisation of both stakes and threats, national leaders have no option but to ensure security, both far and near, by contributing to resolving crises, not with weapons of destruction, but with weapons designed to persuade through controlled strength.

However, circumstances and conflicts evolve very rapidly. In a continuous process of metamorphosis, the threat takes on new forms, for which we are not prepared. Thus, whatever the type of engagement, our forces will be effective only if we adopt a position of observation and adaptation. More so than in the past –but probably less so than in the future– rapid adaptability is proving to be the key characteristic of military systems. Experience shows that the armies that win are those which have learnt well and have been able to put what they have learnt to good use later. "Learn and adapt": this is crucial. We must go much further than where we are today; it is not enough to make note of what we have learnt, we must study these notes and apply them to our force models, the balance of forces, unit training, instruction, etc.

THREE MAJOR TYPES OF ENGAGEMENT

We need to prepare for three major types of intervention.

The first is the short conventional confrontation, in which technology plays an essential role in reducing the opponent's strength. Advanced technology is indispensable since, while war remains essentially a human confrontation, technology reduces human effectiveness. However, although advanced technology is indispensable, it must also be reasonable since, in a time of budgetary constraints, technology competes financially with numbers and, contrary to certain opinions, the inevitable expansion of the battlespace means that numbers remain important; we must ensure that we get the right level of technology, not too much but not too little.

The second type of intervention involves war amidst populations. This will be by far the most common and the most drawn out, since the population will now be at the heart of all interventions, with combat usually taking place in their midst, in an urban environment. This type of intervention –the core of tomorrow's war– marks a return to the truth of war, which is fundamentally a fight for freedom of action. This is the very essence of war. This means that the ultimate aim of war is to "control". You can destroy, you can carry out a precision strike, you can pulverise, you can patrol the high seas, you can overfly a territory for years, you can launch a nuclear attack, but if you do not have "control", this will all be in vain. You may have control of the sea or the air, but if you do not have control on the ground, in among the people, this will all be to no avail. Since the beginning of time, there has been only one way of exercising control, be that at home or abroad, and that is to be present in sufficient numbers in the actual environment in which crises emerge, grow and are resolved: on the ground.

The third type is intervention on behalf of "our population", for their safety and to help them. Such interventions are fundamental, as they are at the very basis of the existence of armed forces, "national protection" being the ultimate aim on which everything else is founded. In this area, land forces, which are able to deploy large numbers of trained and organised personnel at extremely short notice and, thanks to their professionalism and experience of operating in close contact with the population, to improve precarious situations very rapidly, are a vital instrument supporting the state in its essential functions.

As the new century progresses, we see the emergence of a new model. The importance of actions other than combat is now an important characteristic of operations, with rapid shifts between types of action and of conduct. The Cold War soldier, trained to fulfil a specific role, has made way for a multi-tasking successor, who is able to carry out a range of actions based on contrasting skills and conduct, such as coercion, security and humanitarian aid. The commander remains a leader of men, but he must now also be an administrator, a negotiator and a mediator and must, of course, have the knowledge, skills and resources to fulfil these new roles.

The new soldier must understand his place in the regulation of conflicts; it is crucial, but not unique. He must be aware of the importance of the action as a whole and of the fundamental role played by the various non-military actors, drawn from the world of diplomacy, security, humanitarian aid, economics and business. The new soldier must learn to work together with civilian actors to prepare, in advance, the decisive phase of operations, the stabilisation phase. He must learn how to achieve a smooth transition between military and security aspects, between urgent humanitarian relief and reconstruction and development policies. He must also learn how to hand over the baton in the shared journey towards normalisation and how best to combine civilian and military effectiveness to resolve the crisis.

These new roles will not be learned just by training for coercion operations. We now know for certain that the old rule of "he who can do more can do less" no longer applies, either to men or equipment. It is no longer a question of doing less, but rather of doing something different in a different way. The difference is no longer in the level, but in the nature of the action. The spectrum of actions to be carried out by a soldier has become far more diverse, thereby making his life far more difficult; he must remain an expert in the roles of the past, while also excelling in those of tomorrow's war. The military profession has thus widened immensely. It marks, in fact, a return to the reality of the past, reflecting a return to history.

We know that military institutions are particularly good at coming up with impressive solutions to the sort of problems they like to solve, rather than to the problems that will be posed by their future adversaries. The challenges we are facing today mean that we must curb this trend and resist the temptation of continuing to rely on our superiority in terms of technology and destructive capability when dealing with opponents who are very different from those of the past.

We cannot ignore the changes in the world and among our potential adversaries. We have no option but to understand what tomorrow's war will be. By concentrating too much on techniques and capabilities, understandably echoing others around us, we have forgotten to ask ourselves some fundamental questions about the purpose of military engagement. Since it played such an important role in enabling us to fight yesterday's war better, we stuck to the belief that our technical prowess was still suited to the changes in war, without understanding that the face of war has undergone a marked change. The subject of Transformation must be looked at very closely; it is not a matter of discussing technology or organisations, but rather the ultimate aim of war and the best way of achieving the desired result. Should we go as far as instituting a "Counter-Revolution in Military Affairs", to borrow the term used by Ralph Peters?[1] Perhaps not, but we will certainly have to move towards a "transformation of the Transformation", in other words, towards a real ability to understand and defeat the, sometimes very radical, new threats which fall outside what is, wrongly, perceived to be traditional military action. We need to prepare to fight the war we will undoubtedly encounter, not the war we would prefer to fight because it is so familiar; this will mean some far-reaching changes in our way of thinking. With the shifting of the balance towards the perennial truth of the past, war is once again demonstrating that it is not purely a matter of weaponry, but an infinitely more complex and uncertain political, social and human process: we must take control of the excesses of the "digital" culture –and its search for certain and definitive solutions– and ensure that our strategically analysis is not driven by technological creativity.

We need to reintroduce the political dimension into our technical and operational thinking.

It is perhaps because it had missed out on these reflections, that the Israeli intervention in Lebanon was out of tune with reality. Of course, the Israelis did not "lose" but, faced with an asymmetric opponent, the setback suffered by a Western power that does not "win" has significant repercussions. Whatever we may think of the success of our own engagements, we all suffer from the collateral damage caused by all the difficul-

---

1. Counter-Revolution in Military Affairs, Weekly Standard, 6 February 2006.

ties encountered by a Western force: what has changed, in any case, is the other side's perception of the conventional military strength that still shapes us and, consequently, the utility of our current force models.

Over the last few years, we have all seen that it is possible to circumvent conventional military strength. We are thus faced with the urgent need to restore the credibility and effectiveness of our armed forces. We need to reflect on the changes to be made and the new balances to be achieved, and to stop seeking ever more perfect solutions to questions that are no longer being asked.

We need to think otherwise. We need to prepare for tomorrow's war.

BIBLIOGRAPHY

Andréani Jacques, *L'Amérique et nous*, Odile Jacob, Paris, 2000 [America and Us]

Andréani & Hassner, *Justifier la guerre?*, Les Presses de Sciences Po, Paris, 2005 [Justifying War]

Badie Bertrand, *La diplomatie des droits de l'homme : entre éthique et volonté de puissance*, Fayard, Paris, 2002 [The Diplomacy of Human Rights : Between Ethics and a Desire for Power] ; *L'impuissance de la puissance*, Fayard, 2004 [The impotence of power]

Baud Jacques, *La guerre asymétrique*, Editions du Rocher, Monaco, 2003 [Asymmetric Warfare]

Beaufre André, *Introduction à la stratégie*, Hachette, Paris, 1998 [An Introduction to Strategy]

Blin Arnaud, *La terreur démasquée*, Le Cavalier Bleu, Paris, 2006 [Terror Unmasked]

Canto-Sperber Monique, *Le bien, la guerre & la terreur*, Plon, Paris, 2005 [Good, War and Terror]

Challiand Gérard, *L'Amérique en guerre*, Editions du Rocher, Monaco, 2007 [America at War]

Churchill Winston, *My Early Life: A Roving Commission*, Thorton Butterworth, London, 1930

Clausewitz, *On War*, Princeton University Press, Princeton, New Jersey, EU, 1976

Cooper Robert, *The Breaking of Nations: Order and Chaos in the Twenty-First Century*, Atlantic Press, 2003

Coutau-Bégarie Hervé, *Traité de stratégie*, Economica, Paris, 1999 [Treatise on Strategy]

Delmas Philippe, *Le bel avenir de la guerre*, Gallimard, Paris, 1995 [The Glorious Future of War]

Delpech Thérèse, *Savage Century: Back to Barbarism*, Carnegie Endowment, Feb 2007

Desportes Vincent, *L'Amérique en armes : anatomie d'une puissance militaire*, Economica, Paris, 2002 [America in Arms, Anatomy of a Military Power] ; *Deciding in the Dark*, Economica, Paris, 2008

Douhet Giulio, *La maîtrise de l'air*, Economica, Paris, 2007 [Controlling the Air]

Dufour Jean-Louis, *La guerre, la ville et le soldat*, Odile Jacob, Paris, 2002 [War, the City and the Soldier]

Les crises internationales, de Pékin à Bagdad, Complexe, Bruxelles, 2004 [International Crises, from Beijing to Baghdad]

Eisenhower Dwight D., *At Ease: Stories I Tell to Friends*, Doubleday & Company, New York, 1967

Foch Ferdinand, *Les principes de la guerre*, Economica, Paris, 2007 [The Principles of War]

Galula David, *Contre-insurrection, théorie et pratique*, Economica, Paris, 2007 [Counter-insurgency, Theory and Practice]

Gaulle Charles de, *Le fil de l'épée et autres écrits*, Plon, Paris, 1994. [The Edge of the Sword, and other Writings]

Gia Nguyen, Guerre du peuple, armée du peuple, Petite collection Maspero, Paris, 1967 [The People's War, the People's Army]

Gordon & Trainor, *Cobra II: The Inside Story of the Invasion and Occupation of Iraq*, Pantheon Books, New York, 2006

Goya Michel, *Irak, les armées du chaos*, Economica, Paris, 2008 [Iraq, the Armies of Chaos],

Gray Colin S., *Another Bloody Century: Future Warfare*, Weidenfeld & Nicolson (14 Jul 2005)

Grintchentko Michel, *Atlante-Aréthuse, une opération de pacification en Indochine*, Economica, Paris, 2001 [Atlante-Aréthuse : a Pacification Operation in Indochina]

Gros Frédéric, *États de violence*, Gallimard, Paris, 2005 [States of Violence]

Hammes Thomas X., *The Sling and the Stone: On War in the 21st Century*, Zenith Presse, St Paul, Minesota, 2006

Hoop Scheffer Alexandra (de), *Hamlet en Irak*, CNRS Edition, Paris, 2007 [Hamlet in Iraq]

Henrotin Joseph, *La dérive technologique du discours stratégique américain, thèse de doctorat de Sciences politiques*, Université libre de Bruxelles, 2006 [The Technological Drift of American Strategy – PhD thesis]

Howard Michael, *The Invention of Peace, Reflections on War and International Order*, Yale University Press, 2001

Kagan Fredeick W., *Finding the Target*, Encounter Books, New York, 2006

Liddell Hart, B.H., *Strategy*, Penguin Group, New York, 1991

Lynn John A., *Battle*, Westview Press, Boulder, Colorado, États-Unis, 2003

Machiavelli, Le Prince, NRF, Paris, 1968 [The Prince]

Moreau Defarges, *Droits d'ingérence*, Les Presses de Sciences Po, Paris, 2005 [The Rights of Intervention], [

Münkler Herfried, *Les guerres nouvelles*, Alvik, Paris, 2003 [New Wars]

Murray and Millet, *Military Innovation in the Interwar Period*, Cambridge University Press, 1996

Nagl John A., *Learning to Eat Soup with a Knife*, The University of Chicago Press, 2005

Owens Bill, *Lifting The Fog of War*, Farra, Strauss & Giroud, New York, 2000

Paret Peter, *Makers of Modern Strategy*, Princeton University Press, Princeton, New Jersey, Etats-Unis, 1986

Quiao Liang, Wang Xiangsui, *La guerre hors limites*, Editions Rivages, Paris, 1999 [Unrestricted Warfare]

Ricks Thomas E., *Fiasco: The Military Adventure in Iraq*, The Penguin Press, New York, 2006

Rosnay Joël (de), *Le macroscope*, Editions du seuil, Paris, 1975 [The Macroscope]

Schnetzler Bernard, *La guerre de demain*, Economica, Paris, 2004 [Tomorrow's war]

Smith Edward A., *Effects Based Operations*, CCRP, DoD, Etats-Unis, 2002

Smith Rupert, *The Utility of Force,* Penguin Books, Londres, 2005.

Sun Tzu, *The Art of War*, Oxford University Press, Oxford, New York, EU, 1971

Thucydides, *The History of The Peloponnesian Wars*, Penguin, Baltimore, États-Unis, 1972

Tisseron Antonin, *Guerres urbaines : nouveaux métiers, nouveaux soldats*, Economica, Paris, 2007 [Urban Warfare : New Professions, New Soldiers]

Trinquier Roger, *La guerre moderne*, Economica, Paris, 2007 [Modern Warfare]

Van Creveld Martin, *The Transformation of War*, New York, Free Press, 1991

Weigley Russell F., *The American Way of War*, Indiana University Press, Bloomington, 1977

Woodward Bob, *Bush at War*, Simon & Shuster, New York, 2002 ; *State of Denial*, Simon & Shuster, New York, 2006

TABLE OF CONTENTS

Part two

THE NEW CONDITIONS
FOR MILITARY EFFECTIVENESS

The "Stratégies & Doctrines" Series

ARDANT DU PICQ Charles, *Études sur le combat*.

ARMÉE DE TERRE, *Les forces terrestres*.

ARMÉE DE TERRE, *Tactique générale*.

AUDROING Jean-François, *La décision stratégique*.

BECKER Cyrille, *Relire* Principes de la guerre de montagnes *du lieutenant général Pierre-Joseph de Bourcet*.

COURRÈGES Hervé (de), GIVRE Pierre-Joseph et LE NEN Nicolas, *Guerre en montagne – Hier et demain : renouveau tactique*.

DESPORTES Vincent, *La guerre probable – Penser autrement*, 2ᵉ éd.

DESPORTES Vincent, *Comprendre la guerre*, 2ᵉ éd.

DESPORTES Vincent, *Décider dans l'incertitude*, 2ᵉ éd.

DESPORTES Vincent et Phélizon Jean-François, *Introduction à la stratégie*.

DURIEUX Benoît, *Relire* De la guerre *de Clausewitz*.

FOCH Ferdinand (maréchal), *Les Principes de la guerre*.

FORGET Michel, *Puissance aérienne et stratégies*, 2ᵉ éd.

FRANCART Loup, *La guerre du sens – Pourquoi et comment agir dans les champs psychologiques*.

FRANCART Loup, *Livre gris sur la sécurité et la défense*.

FRÉMEAUX Jacques, *Intervention et humanisme – Le style des armées françaises en Afrique au XIXᵉ siècle*.

GALULA David, *Contre-insurrection – Théorie et pratique*.

GOYA Michel, *Irak – Les armées du chaos (2003-2009)*, 2ᵉ éd.

GRAY Colin S., *La guerre au XXIᵉ siècle – Encore du feu et du sang*.

GUIBERT Jacques (de), *Essai général de tactique*, 1772.

GUIBERT Jacques (de), *De la force publique*.

HAÉRI Paul, *De la guerre à la paix – Pacification et stabilisation post-conflit*.

HENROTIN Joseph, *La technologie militaire en question – Le cas américain*.

KARPOV Anatoly et PHÉLIZON Jean-François (propos recueillis par Bachar Kouatly), *Psychologie de la bataille*.

KERDELLANT Christine, *Relire* Le Prince *de Machiavel*.

LA MAISONNEUVE Éric (de), *Stratégie, crise et chaos*.

LE ROY Frédéric, *Stratégie militaire et management stratégique des entreprises*.

MARILLER Roseline, *Quelle stratégie pour l'Europe de la défense ?*

MARTEL André, *Relire Foch au XXIᵉ siècle*.

MONTROUSSIER Laurence, *Éthique et commandement*.

MOUNIER-KUHN Alain, *Chirurgie de guerre – Le cas du Moyen Âge en Occident.*

GÉNÉRAL Palat, *La philosophie de la guerre, d'après Clausewitz.*

PHÉLIZON Jean-François, *L'action stratégique.*

PHÉLIZON Jean-François, *Relire l'*Art de la guerre *de Sun Tzu*, nouvelle édition entièrement revue et corrigée.

PHÉLIZON Jean-François, *Trente-six stratagèmes.*

POAST Paul, *Economie de la guerre.*

ROYAL Benoît, *L'éthique du soldat français – La conviction d'humanité.*

SCIALOM Michel, *La France – Nation maritime ?*

SMITH Rupert, *L'utilité de la force – L'art de la guerre aujourd'hui.*

TISSERON Antonin, *Guerres urbaines – Nouveaux métiers, nouveaux soldats.*

TRINQUIER Roger, *La guerre moderne.*

VENDRYÈS Pierre, *De la probabilité en histoire.*

VILBOUX Nicole, *Prévention ou préemption ? – Un débat d'aujourd'hui.*

YAMANAKA Keiko, *Relire le* Traité *des cinq anneaux de Miyamoto Musashi.*

YAMANAKA Keiko, *Relire* Bushidô – L'Âme du Japon *d'Inazô Nitobe.*

YAKOVLEFF Michel, *Tactique théorique.*

FONDATION SAINT-CYR

The aim of the state-approved Saint-Cyr Foundation is to develop Defense training and research with a view to achieving international academic excellence and cultural cooperation through an enhanced partnership between civil society and the military.

Born in the French Military Academy of Saint-Cyr, after which it was named, the Saint-Cyr Foundation acts for the benefit of all of the French Defense Ministry institutions of higher education within the Army, the Air Force, the Navy and the Gendarmerie. Like many companies, the armed forces are increasingly faced with missions that require a wide range of skills and the capacity to interact with a variety of partners. As an inter-service institution dedicated to pooling skills and aptitudes, the Saint-Cyr Foundation meets the need for common research between civilians and military personnel, between the public and the private sectors, between French people and foreigners. It brings into contact those partners who act together in a very concrete and transparent way to imagine tomorrow's style of command and leadership.

Its actions are carried out along two main lines:
- For the benefit of universities and the Defense Ministry institutions of higher education by funding research and teaching, by providing support for research work, expert seminars and the translation of books and publications.
- For the benefit of businesses and local authorities with a platform devoted to research, training and the sharing of expertise on issues common to civilians and military personnel through specifically tailored training courses, the creation of professorships, conferences and theme-based events.

The Foundation thus provides support in three specific fields of expertise: defense and security studies, dual civil-military research and French strategic thinking with a view to promoting its development and influence. The ambition of the Saint-Cyr Foundation is to enable French Defense research to become a key player at international level and also to become the internationally recognized equivalent of the well-known foreign foundations so as to be able to initiate efficient joint partnerships which will enhance the status of French research.

The support provided for the publication of research work is a major objective which the Saint-Cyr Foundation promotes by awarding financial assistance and also the "Fondation Saint-Cyr" hallmark to foreign language publications and to books and publications by contemporary French strategists and military historians. The goal pursued by the Saint-Cyr Foundation is to boost the development of a genuine French school of strategy and provide the means necessary for it to enhance its prestige and influence.

Fondation Saint-Cyr
Ecole Militaire – 1, place Joffre – 75007 Paris – France
Téléphone : +33 (0)1 45 55 43 56
www.f-sc.org – contact@f-sc.org

Achevé d'imprimer en France en avril 2009
dans les ateliers de Normandie Roto Impression s.a.s.
61250 Lonrai
N° d'impression : 091227
Dépôt légal : avril 2009